Dangerous Experiments

Archeons, book 2

by James L. Steele

Dangerous Experiments (Archeons, book 2)
Copyright © 2019 by James L. Steele

Cover art by **TomTC**, TomTC.deviantart.com
&
Klongi, Twitter.com/klongiarts

Editing by **Renee Carter Hall**, www.reneecarterhall.com

Published by KTM Publishing

Print edition set in Fanwood, Exo, and Playfair Display, all royalty-free typefaces

Print edition ISBN: 978-1-7322824-1-4

Earth

The raptor's voice in Stephen's right ear: "For a normal, mature society to form, at least two intelligent species must coexist on the same world. The two races must be different from one another, and they must not be able to escape one another. This forces them to live together and try to understand each other."

The fox's voice in his left: "This perspective allows for both species to perceive themselves for what they really are, which is only possible by comparing yourself to someone else."

Deka resumed. "Without this, a species will fight amongst itself until it destroys its own kind. They perceive the differences between individuals in exaggerated ways."

Kylac again: "Without a companion species, humans have grown up with the idea that everything that is like you is good, and everything that deviates from that is bad."

"We noticed this pattern in your entertainment. Movies reinforce the impulse constantly, that whatever is not human is evil. Thinking that way helped you survive predators when your species was young."

"But this is not how most species develop. Species with a companion race grow up together. Sometimes they are predator and prey, sometimes they are both predators, and sometimes they are both prey. Regardless, once they realize there is another intelligent species on their planet, they want to learn as much as possible about this other creature."

"Those without a companion race learn that differences must be snuffed out. Anything that does not resemble themselves must be destroyed because that is how they learned to survive. This is why your world is the way it is, Stephen. You are not mentally prepared to meet an alien race. No matter what happens, you will think of them as monsters, and will always regard them with suspicion and dread. It would take a few hundred generations to work this response out of your kind. It is why we generally avoid contact with lone species."

"It's because you are alone, Stephen. Earth has only one sentient species, and this is not normal. Fix that, and everything will eventually correct itself."

Stephen had long ago lost track of who was speaking. Deka and Kylac often finished each other's sentences, one picking up where the other paused to breathe. They were like a married couple to the nth degree.

He was sitting on the couch. Kylac sat on his left, and Deka lay on the floor just off to his right. After the first minute he had stopped looking at who was speaking and just let his ears do all the work.

"Is that it?" Stephen said.

"That is it," Kylac said.

"That's why there are wars? That's why there's racism? That's everything?"

The theropod lying on the floor craned his neck a little closer. In the dim light cast by the one lamp, his dark blue scales blended in with the carpet so well Stephen could barely see him. "If you had to deal with another sentient species on this planet, none of that would have happened."

"Are you sure? We probably would've had to defend ourselves from them. We would still look at them with suspicion."

"You're thinking about it from your current perspective, Stephen," the raptor replied. "What if, thousands of years ago, you knew you weren't alone? Your kind seems inquisitive enough. You would have done everything to learn more about this other species."

Stephen held a hand up. "That's giving us way too much credit. I think we would've destroyed it. We've done it before."

"Believe it or not," Kylac said, "that never happens. As soon as one race learns of another form of intelligence on their planet, everything stops, and they devote their energy to learning more about it."

"It is how a normal planet develops. When a species is alone, it tends to turn in on itself. It never learns how to embrace anything different. It never learns how to experience the world from someone else's point of view. It never discovers portal physics."

The raptor continued. "Only species that learn how to perceive things from the point of view of another can achieve the state of mind required for portal mathematics."

"The key to interstellar travel is having a companion race. This is what has been missing from humanity from the very beginning. This is the hole in your collective minds you have been trying to fill with religion, with quests for wealth and power, with sexual gratification. Without another creature equal to you to share your world, you have turned inward on yourselves, and you are unprepared for the realities of contact with other life."

"You know about religion?" Stephen said.

"Not directly, but your kind would have developed it. You are the correct type."

"What do you mean?"

"One thing at a time, Stephen. Do you understand what is wrong with your planet?"

Stephen shook his head. "No..." He shook it again. "No, if we had another intelligent race on this planet, wouldn't we spend all our time fighting it? Wouldn't we have to fight for resources or something?"

"At first, maybe," Deka said, "but there always comes a point when both sides realize the other is intelligent. They then devote their time to understanding this life."

"They learn how to perceive the world from their perspective," Kylac resumed. "Then they want to help this other species learn how to understand themselves. That is when they mature. Two species, one race. Every species in the contacted universe has been through this. It leads to transcending the limits of their own senses and learning how to perceive the patterns that make up the universe. This allows some of them to open portals through space-time and travel to other planets. They inevitably travel to an inhabited world and join the contacted universe."

"All of this can take centuries or millennia," Deka resumed. "In that time, each has learned from the other. The search for knowledge and understanding becomes central to their lives because it brought both of them higher. That is normal. But when a species is alone, it will continue to wallow in its animal ways. The people will fight amongst themselves for survival because that is all they know."

Kylac picked up the thought exactly where Deka left off. "It worked for their kind's survival for millions of years, so they keep doing it because they have no one else to push them in another direction. Deka and I watched a lot of TV while we waited for you. It's painful, but we wanted to learn as much as we could about you. We have seen much evidence for this."

"The stories you tell consist of lies meant to reinforce false assumptions," said Deka. "The current events on this planet are full of violence but presented for its entertainment value. Sex is censored and yet glorified. You cling to old superstition, and your way of life—no matter how sophisticated it may seem—is still survival of the fittest."

Stephen shook his head. "That's it? We are alone? That's it?"

Deka rubbed the claws on his hands together. "Every problem is simple when you step far enough away."

"But... There's gotta be more to it than that."

"What else do you want?"

"All right, that's the problem. What's the solution?"

"The solution would require time travel." Deka now clicked his claws. He had told Stephen how he laughed so the human would not be disturbed by it anymore.

"Going to the past," Kylac resumed, "and inserting a companion species for humanity to interact with. Think of how many problems you would prevent. Instead of a society of conquest, bloodshed, and bickering, you would devote that energy to learning about the aliens living on your own world. You would learn how they view the world, and they would learn how you perceive it. Your lives would become about the pursuit of knowledge and understanding instead of mere survival and competing for resources."

"A mature civilization develops away from all of that," Deka continued. "Your kind has about fifty thousand years of loneliness imprinted in its psyche. It would take about that much time to undo it now. So... time travel would be the answer, if it existed."

"We're doomed? Is that what you're trying to tell me?"

"Yes," Kylac said.

"That's it? There's no way out?"

"This isn't a movie, Stephen," said Deka. "There is no convenient device written into the story that allows humanity to win despite the odds."

"That's right," Kylac resumed. "Your kind will never make contact with another sentient species, especially trying to sail between the stars as you envision it."

"Eventually, infighting will destroy your kind from within," Deka said, "because nobody is here to show you a different way to live."

The three sat in silence for a while. Stephen was lost in thought. No more questions. It was finally sinking in. The answer. The solution. The root cause of all of Earth's problems. So simple it required enormous brainpower to accept.

"Uh, excuse me," Stephen said, standing up. "I need to take a shower. Maybe lay down."

Stephen walked into the bathroom and closed the door. The sound of running water drifted through the house.

"I hope he's taking it well," Deka said, in Relian.

"I think he'll be fine. He just needs time."

The human had listened with his mouth open, barely breathing, the whole time Deka and Kylac told him they had discovered what destroyed their homeworld. The disaster sent shockwaves across the entire contacted universe, killing many Archeons and incapacitating many more. These were the people who held open the portals between worlds, and without them, civilization itself was on the verge of collapsing. As the only two Archeons who could make offworld spheres, Deka and Kylac had spent over a year traveling to other planets, helping those worlds recover, saving countless lives. Then at last they found Rel's two other Archeons, Rive and Friend. Rive's fox had been experimenting with opening portals outside the universe, into the medium on which the universe itself theoretically existed and moved. Time travel, in effect. Friend apparently could not control the portals he opened into this place, and these antispheres grew out of control, destroying entire solar systems. Rel had merely been the first casualty. Stephen barely seemed to grasp the scope of what had happened. Deka and Kylac still did not fully understand it.

The shower lasted less than five minutes. When Stephen came out, he was dressed only in boxers. He walked into the living room, carrying his other clothes. According to the VCR, it was past three in the morning. Stephen had been home since ten at night.

"Listen guys... It's very late, I just did a seven-hour drive, and my nephew kept me up all week. I'm goin' to bed. Maybe I'll be coherent in the morning. You're not leaving, are you?"

"Not yet," Deka said.

"Good. Please don't go anywhere. I have a lot to ask. I just... I can't think right now. G'night." He walked to his bedroom but did not close the door.

Kylac sighed. "I wonder what that will do to him."

"Hopefully it answers his questions and helps him explain a few things."

"Well, I'm going to hop in bed with him. Maybe he'll be in a good mood in the morning."

"I think you should give him some space tonight."

"I'll let him decide." Kylac stood up, tail lashing. He walked around the corner, down the short hall, and into the bedroom.

Deka raised his muzzle and faced down the hall. He wondered.

2

Stephen felt like a little kid again, sleeping with a stuffed animal. He clutched it, clung to it, squeezed it. It growled, but not in a threatening way, like a dog while happy, but much, much deeper.

Stephen began to wake up. He realized the stuffed animal was warm. With a heartbeat. Stephen opened his eyes and sat up partway. Kylac lay on the other side of the bed, staring at him. Stephen rolled to his back and stared at the ceiling. He noticed he was on top of the covers, and his boxers were missing.

He remembered last night.

"How did you do it again?"

"I didn't do anything." He heard Kylac's tail thumping the mattress. He had told Stephen how he laughed, too, so now the human recognized it. "You grabbed my sheath this time."

Stephen held his forehead. "This is too much. I meet aliens, they tell me we're all doomed, and I start it this time." Stephen took multiple deep breaths. He couldn't seem to get enough air. He turned and looked at Kyalc. "Holy shit, what's happening to me?"

He noticed Deka's head resting on other side of the mattress. The raptor was lying on the floor, staring at him.

"I need to stay away from you, Kylac. I sin against God every time I get near you."

Kylac scooted over to Stephen and lay next to him, one arm across his chest, one leg over Stephen's. "You're anxious again. Remember last time I forbade you from feeling guilty about it."

Stephen remained a surfboard for a moment, and then he breathed again and lowered his arm around Kylac's back.

"You guys were gone for a month. I, uh, I had a bit of a nervous fit."

"What do you mean?"

"I couldn't stop shaking. Running to the bathroom six, eight times a day. Asshole felt so tight it felt like it was trying to hide in my stomach. Ended up going to a doctor. He gave me some anxiety meds. They helped, but..."

Kylac held him tighter, nuzzled his neck with his snout.

"They took care of the physical stuff, but I couldn't stop thinking about that night. I didn't want to face it. That was the best sex I've ever had, and I wanted to do it again. For weeks at a time, I looked in the mirror, and I didn't know who was staring at me."

"Oh, Stephen. You've been forced to ignore this part of yourself your whole life. Now you're facing it."

"At work I caught myself looking at both men and women. Wondering... What's he packing? Looking at asses. Men, women. Both made me hard... Never would have thought it before. This isn't me. This isn't who I am. Now..."

"This is you. It will take some time to accept it."

"I don't know if I want to. I mean, holy shit, last night was great, but... What the hell is happening to me?"

"You have a lot of preconceptions to forget."

Stephen turned to Deka, who lay with his head perched on the edge of the mattress. "You look content."

Deka rubbed his claws together out of Stephen's view.

"What are you doing in here? Thought you were the cold and distant one."

"The carpet in here is slightly less irritating on my scales."

"You just like to watch?" Stephen asked. "You didn't get involved."

"I have a mate."

"Oh."

"And you were in control of Kylac all night."

"Was I?"

"Kylac was just going along with whatever you wanted to do. He needed that after what we just witnessed. And so did you."

The human sighed. "Yeah. I needed that. I wanted it more than anything."

He slid out from under Kylac and off the bed. He stretched, walked to the dresser, and threw on a shirt and boxers. He walked out of the bedroom and into the kitchen. The answering machine was flashing. Three messages. They would all be from work. The clock face showed the time at just past noon, so he was five hours late. He did not care—foolish him for not taking an extra day off to recover from the trip.

Stephen poured himself a bowl of cereal, walked to the couch, and sat in front of a blank television.

Deka and Kylac sat on either side of him, Kylac on the couch, Deka on the carpet. They both scented the bowl from a distance.

"Whose breast milk is that?" Deka asked.

"A cow."

"You drink another animal's milk?"

"Yup."

Deka did not say another word.

Stephen finished his cereal, then stood up and walked to the kitchen. From the living room, Deka and Kylac heard a garbled, machine-generated voice.

"Mr. Penarrow, good morning, it's eight-thirty in the morning. October seventeen. You're supposed to be back to work today. If you could give me a call that would be great. Bye."

...beep...

"Stephen, it's nine o'clock. Just wondering where you are. Please call me back as soon as you get this. Bye."

...beep...

"Stephen, ten o'clock now. Please get back to us. It's been difficult here without you. The other guys have to make up for you. Not a big deal, but I'd like to know what's going on."

...beep...

Stephen turned the corner and leaned on the wall. "That was work. I didn't show up or call off. They want to know where I am."

"What is work?" Kylac said.

"It's where I have to go to make money or I lose my house."

"Ah, monetary social system," Deka said. "We figured that's how this place functioned."

"At least in America." Stephen sighed. "I'm supposed to be at work. I need to go. I don't wanna leave you two here."

"It will be at least a day until we can calculate a way offworld," Deka said. "If you need to do something, do it."

Stephen sighed. "There's so much I want to ask you. So much I want to do. Ugh, but I gotta go. Even if it's just half a day."

He went to the door and slipped on his coat and hat. "I'm really sorry about this, guys, but I can't afford to lose too much money. Not with winter here. I'll be back in a few hours. I promise."

He walked out the door and slammed it shut behind him. Moments later Deka and Kylac heard the car starting, the garage door opening, closing, and then Stephen was gone.

"His reaction is what I expected," Deka said. "We gave him the answer, and nothing changed."

"I think we'll have to push him a little."

"Try sitting closer to him. He has even less self control than you."

"Did you smell the shock when he realized you were in the bedroom, too?" Kylac waved his tail.

Deka rubbed his claws. "He blamed you for starting it!"

3

At six o'clock that night, the doorknob turned, and Stephen finally walked inside. His glasses fogged up, and snow blew in behind him. He was holding bags made of plastic in his gloves. He removed his artificial coat and hat.

"Hey, guys! I brought dinner."

"Where did you go?" Deka said from in front of the couch.

"Did a half shift at work. They wrote me up, but I told them there was a power failure and my alarm clock wasn't working. Didn't think to check because I went to bed as soon as I got back from Philly."

He removed his shoes and placed them by the door. Then he carried the bags to the kitchen. Deka and Kylac followed him and stood where the carpet met the plastic floor. Stephen began emptying the bags. He seemed preoccupied. Several minutes later, Stephen finally spoke.

"They want me to do a double tomorrow to make up for today. I told them yes, I'd do it. Then on the drive home I asked myself... Why? Why did I just agree to that? I got you two here, and you just told me what's wrong with the world. Think you guys can stick around until the weekend? I'll have two days off. I can ask you stuff then..."

"We can't stay forever, Stephen," Deka said.

"I know—I know, I just... If I lose my job, I lose the house. Ain't many jobs around here. Winter's here, and I need to prepare for the heating bill."

"Stephen," Kylac said. "What are you thinking? Whatever you need to say, ask it now. We can't wait for you, and this time we won't be coming back."

Stephen nodded. "Right. All right. Um... On the drive home I was thinking. Thought quite a bit. You're right. We're alone. It's so simple. It's everything that's wrong with the world. But it can't just mean we're doomed. Cold War's over. Nobody's gonna nuke each other anymore. Where are you guys going next?"

"We're not sure," Deka said. "The Archeons are awake. Everyone who can recover from the disasters would have by now. We don't know where Rive and Friend are."

"We'd planned to visit more worlds in case anyone else needs help," continued the fox.

"Please let me come with you."

Deka and Kylac exchanged glances.

"Deka, Kylac, I have no idea what I'm asking, but I won't forgive myself if I don't. I want to see what you're talking about. I want to know what real aliens are. I want to know what we're doing wrong."

Kylac scented him. "You're scared to ask?"

"Yes! I already used up my vacation time, so if I do this I'm gonna lose my job, lose the house—everything! Shit... But what the hell does it matter? You two come here and show me there's a whole universe out there, and what am I doing here? Stuck in a damn factory doing the same shit over and over. Hell with the house and my job. There's a universe out there, and I want to see it! Will you let me come with you?"

Kylac and Deka looked at each other. They knew what they were thinking. Kylac turned back to Stephen and spoke first.

"It's still not safe."

"You two can go anywhere you want. You can make a difference on any world you go to. I can't do any of those things!"

"Coming with us won't change that," said the fox.

"I don't care! You don't know what it's like. I'm in this damn factory and this lonely house. I'm on track to do this until I die, and I can't get out. Nobody can—we all work to survive and then we die. What kind of way is that to spend your life? There's got to be more than this. Please let me see it."

Deka and Kylac did not check with one another. Kylac's ears bloomed.

Deka raised his hand and scraped Stephen lightly down the arm, leaving three tears down his sleeve and raised lines on his skin. Stephen held his arm and laughed nervously.

"What the hell," Kylac said, touching Stephen's cheek.

Deka brought his claws together. "We'll take you on a car ride."

4

Stephen stood in the living room with the Relians, facing the wall. He wore shorts and a t-shirt, definitely out of season here, but Deka and Kylac had told him to dress for a warm climate. He also wore a backpack filled mostly with clothes: long underwear, jacket, jeans, plus a pair of binoculars.

Around his waist he wore a belt. Attached to the belt was a multi-tool—combination knife, scissors, flashlight, screwdrivers, can opener, ruler, chisel, and pliers. He had bought it with the groceries on his way home from work two days ago on a whim, just in case they said yes. It seemed like a good idea to bring it, since he had no claws. He doubted he'd need the screwdrivers or can opener, but the knives, flashlight, and pliers might come in handy.

They had spent the last two days telling Stephen stories of their own. Stories about other planets, how they developed as a society, how portals changed their societies. So many planets, so much history. The phone kept ringing, and it was always work. Stephen almost picked up and told them he was quitting.

Ten minutes ago, he called his sister. He hadn't said much to Melissa, only that he would disappear for a while, he didn't know when he would be back, but not to worry because he was with good people. Melissa had been, of course, exceptionally worried, so Stephen did something he did not expect. He told the truth.

He told her about Deka and Kylac. He told her everything they had told him, as best as he could summarize it. He told her what was wrong with humanity, and that he was going with them to see it firsthand. She had not been convinced. Stephen had said goodbye one last time, and hung up.

The phone was ringing now. It had been ringing non-stop since he hung up. Pretty soon the tape would be full, and Stephen didn't know what the answering machine would do.

"Remember, Stephen," Kylac said, "you are the first of your kind any of these people have meet. They will be curious about you. Be a good ambassador for your species."

"I can't believe you just said that. And meant it."

"This is serious," Deka said. "We're taking you to people who will take your body apart, tell us how it works, and put you back together better than before."

"Really?"

"Oh, yes. You are a brand new species. They will want to know everything about you, and they will preserve your anatomical records as long as they're alive. Just let them do what they need to do, and you will be fine."

"Okay." Stephen's heart was racing. "I have no idea what I'm getting into, but I trust you. I'm ready."

Deka allowed the equations to return their results and lock in place, and his mind reached into reality and connected two distant points in spacetime. A sphere opened. An alien landscape became visible through it. Saber-toothed felines walked about on a grassland. Some saw the portal open, and they approached it. They appeared to be staring at Stephen. A crowd of four-legged cats with long teeth formed, smaller doglike creatures among them.

Kylac nudged the human. "Look familiar?"

"You weren't kidding. They look just like saber-toothed tigers. And those things are like wolves. They're intelligent?"

"They are."

Deka backed up and stood beside the human, leaning his snout next to Stephen's ear. "It's custom for the one who opened the portal to go first, but this time I want you to take the first step."

Stephen faced Deka, then turned back to the portal. He squared his shoulders, took a step, and walked straight through the sphere. The Archeons followed, and the way closed behind them.

The heat in the house was set to fifty degrees to keep the pipes from freezing. The lights were off. The tapes were stacked on the floor. The phone was still ringing.

Selta

I

Stephen opened his eyes. He was vaguely aware he was lying on something that resembled grass, but softer. Strange smells surrounded him. Animal smells. Stephen sat up and winced.

His back was sore. His neck was sore. His chest was sore. He looked down, realized he had no clothes on, and covered himself with a hand. Then he noticed the red line of raised skin running from his crotch to his neck. He held his arms up, turning them over to follow the line twisting around his forearms all the way up to his shoulder.

Saber-toothed felines approached him, making a long series of different-pitched growls. Everything in him wanted to get up and run, but then paws reached behind him and held him down. Felines closed in and formed a ring around him, paws feeling his forehead, tongues licking him in multiple places, growling amongst each other. Stephen's heart raced. They looked like hungry animals getting a taste for him, and he tried to stand up, but they continued to hold him. Finally they backed away and examined him from a distance. Stephen now stood. His clothes were nowhere to be seen, and Stephen didn't recognize anyone in this group.

The cats stood as tall as his chest. The dogs among them were smaller, but still huge. They were growling to each other, staring at Stephen, and the noises passing back and forth between the people were unnerving.

Aliens. Cats and dogs, but aliens. Teeth everywhere. Everyone was growling and snarling, and yet nobody had bitten him.

Stephen slowly remembered how he got here. He had just stepped through the portal, marveling at the fact that he was the first human to set foot on an alien world. A few seconds after he touched down, the sphere closed behind him, and the cats came pouring through the other portals around him.

Kylac had muttered something about word getting around quick when they heard a new species showed up in the contacted universe. Deka had told him to relax, so Stephen did. Then the cats were all over him. Some of the dogs, too. Noses and paws and tongues touched every part of him. It wasn't uncomfortable, and Kylac and Deka had warned him about it, but living it was far different. They pushed him over, touched him in just the right place, and the world fell away.

Now here he was staring them in the face, but completely naked. They were talking. They were saying things. Stephen was sure they were talking about him.

He raised his hand. "Hi. Nice to meet you."

"Nice to meet you, too, Stephen," said a voice behind him.

Stephen spun around. A cat was coming toward him through the pack of giant animals. It halted, standing too close for comfort.

"You speak English?"

"Deka and Kylac taught me while you were unconscious. I'm Sere, one of Selta's Archeons. The other is off-world."

"Hello..." Stephen caught his breath. "You speak it well. Almost no accent."

"Yours is one of the easier languages. Simple sentence structure. Very few inflections. Tip of the tongue. The prepositions are a bit slippery, though."

Stephen gestured to the scars covering his body. "What happened to me? How long was I out?"

"It's been two days. We examined you."

"Ex—?" Stephen inhaled but no air reached his lungs. He tried again. "It looks like you dissected me."

"That is how we examine."

Stephen stumbled back a few steps only to bump into the nose of some other feline. He felt a tongue on his ass, gasped, and hopped forward a step.

"You... You cut me open and poked around?"

"Whenever a new species is contacted, it's common to bring them here first so we can learn about their anatomy. We then spread that knowledge to every planet in the contacted universe. Everyone needs to know what your kind needs to survive, what your limits are, and so forth."

"Yeah. Sure." Stephen looked around the ground. He saw nothing but giant cat paws. "Where are my clothes?"

"With Deka and Kylac. They're sleeping."

Stephen turned. Feline and canine eyes were staring at him.

"Is something wrong?" Sere asked. "You smell terrified."

"Well, yeah. You just dissected me. Where? How?"

"Right where you're standing, and several people did. With this." The cat held up a paw and extended one long, sharp claw. "We tasted your fluids, measured your body proportions, examined your skeleton. We know everything about you."

"And—" Stephen realized he wasn't wearing his glasses. He felt his face. Looked at his hand, then at Sere.

"Normally we would have put you back together as we found you, but we decided to make one change for your benefit."

"You fixed my eyes?" He held his hand far away and brought it closer. It remained in focus the whole time. "How the hell did you do that?"

"We reshaped them for you. The lenses you were wearing were clever, but hardly a solution. Does your species not correct genetic drift?"

"Correct what?"

By now the crowd had begun to disperse. Stephen felt more comfortable without the audience.

"Natural selection is not killing off stray genes," said the feline, "such as unfocused eyesight, and the trait is free to multiply. Instead of correcting the problem, you wear lenses. Why did physicians on your world not correct the eyes?"

"I can't afford laser surgery."

Sere regarded him silently for a moment. "I don't understand what you mean. There are numerous parts of you that are defective in one way or another. Your eyes were out of focus. Your jaw has too many teeth. Your testicles are prone to rotating. Many of your joints show signs of swelling. All of these are easily corrected, and yet they persist."

"I don't think we're as good at medicine as you are."

"Deka and Kylac told me about your planet. Still uncontacted, but they are taking you with them because you wanted to know what was beyond your world. I hope you learn something that will help you."

Sere turned and began to walk away. Seeing this cat from behind, Stephen realized Sere was male. He lost his breath again. The voice had sounded so feminine.

"Wait!" Stephen ran to catch up, still covering himself. "Where are they?"

Sere looked back at him. His face seemed to be locked in a permanent scowl, but he could have been smiling for all Stephen knew. "They will find you when you're ready."

"Wha...?"

"It's your first time offworld. See it. Smell it. Touch it."

He walked away. Stephen stood there, still covering his crotch. He looked around and realized he was in the middle of a field of cats and dogs. There was much growling among them. All of them regarded him as they walked by, but none disturbed him. Several of the canines approached, scented him, grumbled or growled, and walked on.

Nothing but open grassland with a few trees as far as he could see. The light seemed wrong somehow, and Stephen looked to the sky and saw two suns, one white, the other red.

Spheres lined a path not too far away, and many cats and dogs were exiting and entering them, every canine flanked by at least two felines.

Something in the distance caught his attention. The color red. It looked like someone had been killed. Stephen began walking toward it, quickly at first, but stepping quietly and slowly as he neared. Seven cats and one dog huddled around a dog lying on its back, splayed open. The group seemed to have their heads inside the body cavity, licking. Some had their paws in it and were digging around on the inside.

He craned his neck to see over everyone. Several turned and saw Stephen approaching. Stephen jumped backwards a few steps, and one of the dogs backed out of the group and took a few steps toward him, tail raised. Stephen backed away.

Then the dog sat, regarded Stephen, and cleared his throat. "Don't be afraid."

The canine couldn't seem to form the letter *b*, so he had dropped it entirely and extended the *e*. Stephen smiled. "You speak English, too?"

"Few words. New language travels fast. Please see." No *p* sound, so he had replaced it with an extended *l*. The dog stood and turned halfway around.

The strange accent sounded so fascinating Stephen forgot why he came all this way for a moment. Then he saw the red color again and took a step forward. None of the cats surrounding their kill reacted. He approached cautiously until he stood directly behind one of the cats. The cat turned around, enormous teeth glaring at Stephen more than its eyes did, scooched over, and made a place for the human. Stephen slowly knelt into the gap.

The canine, female, had been completely splayed open. All of her insides were visible. The felines were tasting her, running delicate claws through her. They were removing a lumpy, yellow mass from her body. A tumor, or maybe an infection.

They were still digging it out, whatever it was, all these cats tearing through the dog's insides, growling, snarling, yapping. This seemed more like an operation than a feeding frenzy. Different cats were tasting various fluids. Others held their paws to the canine's head. Still others kept touching their claws to certain points on the skin over and over.

Then at once they began licking the insides of the skin, the edges of the skin, the fur on top of the skin, and finally closed the skin over the body, manipulating it with their claws and teeth, holding it together in their mouths up the entire incision. They held still for several minutes, then let go in unison. Finally they stepped back, and the canine now started breathing. Stephen had not closed his mouth since he took a front-row seat.

The group of felines dispersed until only two remained. Stephen tried to follow them, but they were quickly lost in the field of milling cats and dogs.

He spotted the dog who had spoken to him earlier, and Stephen ran to catch up.

"What just happened? What was that?"

The dog tilted his head. Stephen hoped that meant he was thinking.

"They saved the life."

"Was that... Was that surgery?"

"Yes. Disease remove."

"Out in the open, in the grass, with their bare hands?"

A long silence. The canine tilted his head twice. "I understand not."

"I'm sorry," Stephen said, backing away. "Nice to meet you. Really nice." He looked around.

His eyes settled on the canine patient again. Two more felines, different people than before, now lay beside her. One of them was feeling her skull with a paw. The other looked like she was licking the dog's breath.

Stephen walked away. He observed many people, some just sitting in place, seemingly staring at nothing. They followed him with their eyes as he walked by. Many dogs and cats sat or lay down, staring off into space. Many felines examined various dogs, and the dogs seemed to welcome the frequent inspections.

As Stephen walked among the dense gathering of cats and dogs, one of the dogs was rolling over so the cats could sniff him. They tasted urine and feces. Stephen winced but forced himself to look. Several cats jumped in, touching the dog's skin in several places at once. The dog seemed to die right then: motion stopped, breathing stopped. Then the cats raised claws, cut open his skull, and began removing it.

Again Stephen approached, and again they made space for him to sit. He tried not to exhale in the dog's direction as the cats removed the dog's skull. He'd only had some medic training in the Army, but even a field medic could recognize what belonged and what didn't belong, and this orange stuff they were pulling out certainly did not belong on the brain.

The cats peeled off this film, replaced the skull, sealed everything shut with their saliva, and parted ways. The cats removed their hands from the dog, and he started breathing again.

Stephen stood up and stumbled away, feeling dizzy.

2

Stephen wandered into the midst of the spheres on the hub. Kylac and Deka had told him there would be portals but none led offworld, so he would be safe walking through any one of them.

Many saber-toothed cats and giant dogs entered and exited glass balls resting less than an inch off the ground, large enough for someone his size to walk through. They seemed like they should be making a noise—a hum, a whine, a vibration—anything, but they were soundless, and each showed him a view of the other side stretched around it, as if the entire planet were contained inside.

From sphere to sphere, the view was more or less the same. Either the entire planet was grassland, or they only lived on the grasslands. A few portals showed forest, and it was on these portals Stephen saw the most canines. Stephen felt more comfortable around them, as they had smaller teeth, and they seemed to be exclusively on the receiving end of the surgeries.

Stephen chose a portal and walked through, emerging in a forest. It was cooler here but still tolerable while naked. Nobody seemed to notice he had no clothes on. He couldn't remember when he had lost his self-consciousness.

Some of the wolves here noticed Stephen's arrival and approached him. They sniffed him, and Stephen was tempted to pet them, but that felt rude. After they scented him, they growled and grumbled and made all sorts of sounds from the back of the throat, and then walked away as a pack. Stephen hoped that was a good sign. He followed from what felt like a respectful distance.

Stephen's feet could barely take walking on the under-brush and over fallen branches. This forest was almost disappointing in how earthly it appeared, and for a while Stephen forgot he wasn't on Earth.

The wolves started running, and they pulled away like drag racers, so fast Stephen didn't even try to catch up. He watched through the thin trees as something moved so quickly it took this team of twenty dogs to bring it down.

The wolves dragged it in Stephen's direction. The human remained a respectful distance away and observed. Their prey resembled nothing on Earth. It had fur, it walked on four legs, but it seemed to have one and a half heads: the main head at the front, and another head with only eyes on the rear.

Something else zipped by, too fast for Stephen to see, and some of the wolves broke off and gave chase, accelerating to the same blurring speed.

The group with this corpse carried it all the way up to one of the portals. Stephen followed through the woods, cursing and wincing every step, and hopped through the portal the group had entered.

Saber-toothed cats were examining the pack that had returned with a kill. The cats passed up most of the dogs, but one of them was obviously in pain. The cats put her to sleep, opened her up, set a broken rib, closed her back up again, and left her to wake on her own. Others had gathered around the kill and were eating. Stephen sat down and caught his breath. He couldn't shake the feeling that none of this was real, that it was some documentary on television, and he was merely the camera capturing it.

3

Deka and Kylac sat at the hub as the daytime stars slowly sank below the horizon. Sere lay nearby, head on the ground, in a shallow slumber.

"I hope he doesn't hurt himself," Deka said in English.

Kylac's tail was thumping the ground. "If he does, this is the best place to do it. If he falls off a ledge and dies, someone will see it and put him back together again. Besides, he has to come back. We have his clothes."

"He's been gone all day."

"Maybe he found a mate."

"I wouldn't laugh," Deka said. "Stranger things have happened."

"You're right. I shouldn't joke about it.. But can you imagine? First time offworld, and what happens?"

"I hope he's smarter than that. I trust the Selts and the Zjr wouldn't hurt him, but he's supposed to be an ambassador to his species. He might give them the wrong idea about his people."

Kylac rolled over, tail still flipping back and forth. "That would be perfect! Think about it! Stephen wanted to know what else is out there! This is what you've been missing, human!"

Deka rubbed his claws. "Do you think we let him wander around by himself for too long?"

"I don't think there is a too long. It's his first time being off Earth. May as well be Selta. It will remind him of Earth but just different enough to make him feel he's not at home."

"I was afraid he'd be too scared to leave the hub."

"Oh, Stephen's better than that. He just doesn't know it yet. Once he gets over the self-doubt and confusion his society pushed onto him, he'll be fine. He's better already after a couple nights with me!"

"I'm curious to know what Selta is from his point of view."

Sere suddenly raised his head, fully awake. "You didn't push him too far, did you?"

Kylac's tail waved. "Just a compatibility experiment. He was even more compatible than I thought. After his first climax, he said he'd never go back to humans again."

Deka clicked his claws now, changing from a smile to a laugh.

"Do you plan to take him with you while you search for Rive and Friend?" Sere asked.

"We don't know," said the fox. "He wanted to know what's out there, we can show him, so we'll let him come along."

"As long as it's safe," Deka said. "Then we'll send him back."

"Do you have any idea where Rive and Friend might be now?"

Breaths later, they smelled distinctive human scent. All three watched as Stephen weaved between various Selts and Zjr. Some of the cats approached him, bumped his side. Stephen said hello and waved to them as he walked on.

He stopped just in front of the Relians and Sere, acting as though he had an infinite number of words on his tongue. Then Stephen dropped to his knees, looked at all three of them in turn.

"Where are all the buildings? Where are the cities covering the entire planet, the roads, the flying cars? You said this was an advanced civilization, but this is stone age!"

Deka laughed with his hands. Kylac laughed with his tail.

"And how do they do it?" continued the human. "Just drop right there and perform open heart surgery. Total strangers just walk up, cut them open, and leave! What am I looking at?"

Kylac tossed Stephen's clothes to him, then his backpack. Stephen slipped his shirt on.

"And why did I have to go naked? Where the hell were you three?"

Deka stood and wrapped his neck around Stephen's. "Welcome to the contacted universe." He backed up, giving Stephen space to dress.

The human stared at him. "This is normal?"

Deka was stroking his claws slowly and rhythmically. "We had to take your clothes away. Otherwise no one would have felt comfortable around you."

Stephen had just finished putting on his boxers. "Why?"

"You would stand out. Nobody wears clothing."

"Really?"

"The better question to ask is why do you wear it?"

"Because I don't have fur and I'm cold."

"It's warm on this world," Kylac said. "You still wore them."

"It would've been rude." He slipped his shorts on.

"No, we're talking about the deeper reason you wear clothing, Stephen."

"What deeper reason?"

"Your lack of fur is only the beginning," Kylac said. "In primitive times, your species needed to take other animals' fur in order to stay warm. Because of this, you developed a core mindset of taking from others what you do not possess."

Deka finished Kylac's thought. "You live with this even now. It is not your fault, of course. You had nothing to do with it, and you cannot help what you are, but every species you meet will regard you with suspicion. They recognize a species who survives by taking what someone else has and using it to make itself stronger. If you weren't a lone species, they wouldn't mind because you'd have a companion race to balance you out, but since you don't, they know the impulse is uncontrolled. Be prepared to go without clothes frequently."

Stephen stared at them.

Deka spoke again, holding his claws together. "And yes. Spaceships. Technology. Machines. You assume the end result of civilization is technology. Think about it, Stephen."

"Why did someone invent artificial light?" Kylac continued. "To help you see in the dark, which you cannot do on your own. What about the car? To move you around faster than your legs can take you. It's the same reason you wear clothes. These things make up for shortcomings of your own body, and you assume all species must do this, too. But look at the Selts and the Zjr."

"They are mobile enough to go anywhere," Deka resumed. "Strong enough to carry anything they need. The Zjr are able-bodied enough to hunt for themselves, and for the Selts. The felines are evolved for medicine. Portal physics takes them anywhere they need to go. Anything they can't do, they know people of other species who can. If they need somebody who can see into the x-ray side of the spectrum, they find a person who can. They discovered portal physics long before they needed machines, so it never occurred to them to build them. That's what having a companion race does to a species' development."

Stephen spread his arms. "And still no spaceships?"

Kylac's ears flicked. "No, Stephen. Nobody in the contacted universe needs spaceships. We can go anywhere we want simply by being aware of the patterns that make up the universe."

Stephen looked from side to side and waved an arm at the the canines and felines around them. "I've been walking by these people all day. They're just lazing around."

"Actually," Sere said, "they are teaching each other mathematics, science, history, and philosophy. Others are investigating the world around them. Those two over there are eating the grass, learning how to analyze the molecular structure. Deka and Kylac tell me the phrase is teacher and student."

"Where are the books? Where are the pencils to write things down?"

"We remember what we are taught," said Sere.

"Yeah, but you need to write this stuff down, don't you?"

"We have no written language," Sere said. "We speak to each other. We learn what everyone else knows."

"There are only three thousand Selts, and another two thousand Zjr," said Deka.

"Five thousand people? That's it? Jeez, there are five billion people on Earth! Why so few people here?"

"Why so many on Earth?" Sere asked.

Stephen stared blankly at him.

"The larger anything is, the more difficult it becomes to keep it orderly," said the feline. "When the population is so large, you can't know everyone. Society begins to control you, and it turns into a system that is beyond anyone's control. Animals reproduce without boundaries because it's their nature to do so. Reproduction is largely unnecessary once you achieve intelligence."

Stephen turned around, scanning the Celts and Zjr and the grass and the sparse trees,, reexamining everything with this new information. Again Deka and Kylac tried to see the world through the human's eyes. They guessed he was seeing the Selts and the Zjr for the first time now.

"The Zjr are the only ones who can hunt the animals on this planet," the fox said. "The Zjr hunt for two years of their lives, then their hormones calm down and they can begin developing their minds. They still hunt, but not so intensely, and on their terms. They'll keep hunting for the Selts, and the Selts continue healing the canines for their entire lives."

Stephen sighed. "And this is normal?"

"Yes, Stephen," said Sere. "This is normal."

He turned and took in the sights, nearly making himself dizzy. He stopped, facing the setting stars on the horizon.

"Wow."

Stephen sat down and watched. The Archeons watched it with him. They realized it was the first time he had watched two daytime stars setting in the night sky.

Faii

The portal closed behind them. It was a warm place, and everything looked purple to him. The ground was a shade of lilac, the sky was violet, the vegetation was lavender. He looked up at the sky, at the enormous sun, maybe five times the size of Earth's, but dimmer, barely glowing at all. It made everything look purple, even his skin.

"That's weird," he said.

"This is Faii," Deka said. "The star that makes this solar system was never very bright to begin with. It's still boiling hot on the surface, but most of the heat never reaches this world. Faii is a planet fueled entirely by internal processes. It doesn't need its star at all."

"Cool."

There were no trees or grass in sight, just rolling hills covered in vine-like plants. Stephen realized he was standing on a woven surface made of these plants.

"So how do the plants live? Are they everywhere?"

The Relians began walking. Stephen jogged to catch up , looking over his shoulder at the hub and the few other spheres leading to more purple lands.

"The plants cover the entire planet except for the poles," Deka said. "Without them, there is no surface to this world."

"There's no ground underneath?"

"None above the surface," said the raptor. "This planet is covered completely in liquid water. The plants evolved a means of survival by clumping together into large islands. This calmed the constant winds and violent storms that swept over this world. In time, the various islands connected and formed a carpet so thick they became solid enough to live on."

People became visible in the distance. Bipeds and quadrupeds. Stephen squinted, but every time he tried to see something from a distance, it seemed to disappear. He turned to Deka.

"Guys, is this safe? I mean, for everyone to hop from planet to planet like this? What about invasive animals and plants? What about disease?"

"That's why we took you to meet the Selts," Deka said. "They know what you're carrying, its genetic code, what it affects, and what affects you. We know which planets you should not visit."

"How can they be sure? Odds are something will start a plague somewhere."

"You assume every species is a walking colony of microbes and bacteria," said the fox. "But most species develop a much stronger immune system than yours did. Your body coexists with many microbes. A normal immune system kills all intruders, so your diseases won't destroy any planets."

"As for invasive species," Deka continued, "the Selts tell us which will be harmful for which planets, diseases one species carries that can affect people on another planet, and we make sure it doesn't happen. We raise our children not to go there, we place portals in regions where animals can't wander in, but most animals avoid portals, so that is rarely a problem. Many microorganisms are common throughout the universe, so there's no harm in transplanting them."

"Are you telling me aliens don't get sick?"

"Of course they do," Kylac answered, "but the chances of a disease crossing species lines are very slim. Our populations are small enough that disease is usually not a problem, and even if a microbe did jump to a new species, they evolved much less tolerant immune systems compared to yours. Not everyone is as vulnerable to microbes as you. It's actually you we must be careful of."

"It's one reason we brought you here next," said the theropod. "You're about to meet the Ori. They're reptiles. Their companion race is the Heu, and they may surprise you."

"How?"

"We want you to be surprised," continued Kylac, wagging his tail. "Get ready to take your clothes off."

"Again? You're sure people won't see clothes as a human virtue or something?"

"You're part of a lone species, Stephen," Deka said. "They'll know it is no virtue, but an expression of your carnal nature. We're getting close. Now, Stephen."

The human stopped, took off his backpack, and slipped his shirt over his head. He took off his shorts next and boxers last and then stuffed everything into his bag.

"Can I at least keep the shoes?"

"Everything must go," said Kylac.

"Damn. Well, the ground feels pretty soft. Won't be too bad."

It was incredibly cramped in this backpack. He raised it over one shoulder and started walking again.

The people were coming into view now. The reptiles were easy enough to see, but the people next to them...

"When you meet the Heu, remember this is a planet that evolved without soil," Deka said. "This means there is very little bacteria, and very few microbes. Every species ate the plants, but the predators died off on their own. This is probably the cleanest planet you will ever visit. Don't worry about any germs you have contaminating it or killing off the people. Nothing you have will affect anyone here, and they have nothing that can make you sick."

Stephen barely heard him. He was staring at the Heu. When they were close enough to make out, he gasped and halted in his tracks.

"Oh my God. Where is their skin?"

Deka's hands clasped, claws clicking. Kylac's tail twitched. The fox raised a hand to Stephen's back and urged him forward. Stephen moved only as fast as Kylac.

Deka said something in another language. The reptiles were quadrupeds, built low to the ground, resembling monitor lizards, but much larger. They were the only ones making sounds in their direction, apparently answering Deka.

The Heu... They were bipedal, on backwards legs similar to Kylac's, but their muscles were visible. Their internal organs were visible. Stephen could see parts of their skulls. No genitals. Kylac began making broad gestures with his head and arms. All of the Heu gestured back.

Finally everyone came face to face with Stephen. The reptiles tasted his skin with long, forked tongues. Though they were only knee-high, they were twice as long as Stephen was tall.

The skinless Heu were half again as tall as Stephen, and they stared down at him, heads covered in eyes that never blinked. No eyelids, just black orbs embedded into the exposed skull. Stephen counted nine on the side facing him, the side with the mouth. Other eyes faced the other direction, and Stephen saw the back side of their heads were covered in eyes as well.

They moved in broad, exaggerated motions, reminding Stephen of birds. Now up close, Stephen noticed their bodies were coated in a clear slime. He dared not touch it. He didn't even want to breathe on them, as they looked so thin the slightest movement would break them.

Kylac gestured. The Heu gestured back, arms flailing everywhere, fingers waving in different directions, legs moving about.

Deka, meanwhile, grunted and hiccupped to the reptiles on the ground, who did the same in reply.

Stephen had never felt more alone in his life.

2

"What the hell is going on?" Stephen said.

He sat with Deka, Kylac, one of the lizards, and one of the Heu. The soft plant ground was comfortable but made him feel like he would fall through and the planet would swallow him if he sat here too long.

"I want you to meet Bols." Deka gestured with his snout to the reptile standing before them. "She's an Archeon. And we call her Kuu." He pointed at the Heu. "Her name is nonvocal, so offworlders have to make up a name."

"Hi. Please tell me why she doesn't have skin."

"Keep talking, Stephen. They're learning English. I'll let her explain in a minute."

"Umm... right. Well, this is quite a place. Everything is so dim here."

Deka smiled with his claws. "When Kylac and I first came here, we were only children. Kylac heard stories of the Heu, and he had to see them for himself, so Kylac dragged me through a few portals around Rel's hub and finally we arrived on Faii. Go ahead, Kylac. Tell him what you did." He clicked his claws together loudly, laughing now.

Kylac's eyes were closed, as if in pain, but his tail waved about so he was laughing, too. "I was so hungry. The plants smelled so good. You can't smell it, Stephen, but they're the source of oxygen on this world. They draw nutrients up from the depths of ocean underneath, from the volcanic vents that ring this world, so they smell like the ocean and everything in it. I didn't know how to disregard what my nose was telling me and see with my eyes. I smelled fish. I tasted fish in the air. I thought we were walking on fish."

"Kylac ate the plants," Deka said. "He gorged himself on them. I told him they're not fish, don't eat them. He kept saying yes they are, yes they are. Problem is the plants contain arsenic."

"What?" Stephen looked down. "I'm sitting on arsenic?"

"It's completely contained unless you eat the plants. I'll let the fox tell the rest."

Stephen turned to Kylac, who was covering his eyes and nose with one padded hand. "Some of the natives saw me eating the plants. One of the reptiles stuck his tongue down my throat and pulled everything up. He did that for over an hour." He whined.

"Why didn't you just puke?"

"I wanted to, but the reflex was gone. The plants evolved like that. The fastest way to make sure none entered my system was to pull it back out. One tongueful at a time..." Kylac whimpered. "That was the day I learned to use my other senses. Sometimes the nose is wrong."

"That's horrible."

"It saved his life," said the female Ori lying in front of them.

Stephen turned to her and stared. It was as if someone's pet monitor lizard lying in the living room just up and decided to join the conversation after staring at him for hours.

The Heu flailed her arms and kicked her legs while waving her fingers.

"Kuu says she wished she had been there," Deka said.

Stephen did not want to look at her.

"Introduce yourself, Stephen," said Deka. "They can understand you. Kuu has no vocal cords, so we will speak for her."

Stephen turned to the reptile. "Hi. I'm Stephen." Then he looked at Kuu. "My name is Stephen."

She stood up, started moving her body in various exaggerated ballet poses. It was like watching one of those skinless classroom mannequins dance, every muscle and ligament visible as it stretched and flexed.

Kylac spoke for her. "I am Kuu, one of Faii's Archeons. Deka and Kylac have already told me this is your first time offworld and you are an isolated species. I can tell you are surprised by my appearance. Most are. My species evolved codependence on the plants here. My body absorbs the arsenic and distributes it across my body, keeping bacteria out. The Heu never had to evolve skin."

When she stopped, she was facing away from Stephen, the eyes on the side of her head still looking at him, unblinking. It was the most grotesque thing Stephen had ever seen.

"But... Arsenic? Isn't that pure poison to everything?"

She danced again. Kylac narrated. "Not if your body knows how to use it. I understand it is poison for most species. You should not touch me."

Stephen tried not to shudder. "I won't."

3

The Heu were dancing, all fifty of them, and a raptor and a fox danced among them, two animals prancing with sentient, skinless corpses. Disturbing, but also beautiful in a way. Stephen knew this was no dance for entertainment, but an orgy of conversation—a mass exchange of ideas, and the Relians spoke it as well as they spoke English.

They danced for what must have been hours. Stephen was hungry, but he dared not eat the plants. He did not even want to be touching them right now. His ass was sitting on these things, and he hoped his skin wasn't absorbing anything it shouldn't.

Stephen felt bored, and yet he could not take his eyes away from the dance. So much conversation going on, and yet not a single sound. He thought a Selt surgery looked alien, but they had nothing on the Heu. Finally Deka and Kylac bowed out and approached Stephen.

"What were you talking about?" Stephen asked.

"We were telling them about Rive and Friend, and why we've been losing planets," Deka said, sitting down on Stephen's left.

Kylac took a seat on Stephen's right. "A few others were discussing the quantum world."

"The what?"

"The Heu don't converse as we normally think of it," said Kylac. "We are Archeons, so we can focus on many people speaking at once. You, however, can focus on one, maybe two people speaking at a time. The Heu can converse with hundreds of people at a time. Their language is exclusively body motion, so everything they do is speech."

"How the hell do you know all these languages?"

"Archeons learn all of them," Deka said.

Kylac picked up the thought and ran with it. "In addition, we also learn each species' history, plants and animals we can eat, survival on their world, the location of all points to create a portal, the location of each world's place in the universe, orbit speed and shape, rotational speed, tilt—"

"Shit," Stephen said. "Nobody can memorize all that."

"It's part of being an Archeon," Deka said. "The mind is capable of much more than you realize. It is possible to learn all these languages, maintain dozens of portals across the universe, and have a normal life at the same time."

"Oh my God, I have a headache. Just watching you was information overload, and these people... This is kinda gross."

"And yet they're alive and intelligent." Kylac's tail waved. He slapped Stephen on the back.

"Don't give up yet, Stephen," Deka said. "We brought you here to show you something even more interesting."

"Even more?"

"This planet has a surprise, and it's up there."

Deka pointed at the sun, large and dim in the permanent twilight. Stephen looked at it, but nothing seemed special.

"Remember your binoculars?" Kylac said. "Remember you asked us why we wanted you to bring them?"

"Uh huh."

"Take them out."

Stephen unzipped his backpack. Back on Earth, Stephen had wanted to bring a first aid kid, some water, a tape player, and a camera, but Deka and Kylac said no to all of them. They wouldn't even let him bring a towel. All they allowed was the multi-tool, but that seemed more of a burden than anything since he couldn't wear clothes half the time, and yet they had wanted him to bring this bulky pair of binoculars.

He pulled them out, turned back to the sun, and held them up to his eyes. The star already filled up most of the daytime sky, and the binoculars magnified it so much he could see the ripples in the star's surface.

He focused.

Stephen stood up, trying to get a closer look. He took them out of focus, rolled the view back into focus again. He heard Deka smiling.

"What the hell is that?"

Things were on the star. Things that seemed to be swimming. They formed groups, broke off, swam in other directions, joined with others, and moved about once more.

"Is that...? Are they alive?"

"Those, Stephen, are the Faii," Deka said. "They are alive."

Stephen lowered the binoculars and faced his companions, mouth agape. The raptor and the fox were indeed smiling.

4

They looked like dolphins. Stephen had been stargazing for so long he could make out their form clearly now. They had bottlenoses, fins, and the overall shape one would expect a species of aquatic animal to have, but these lived in a boiling hot star—cold by stellar standards, so the Relians had told him, but still a living hell for every other life form in the universe.

Stardolphins.

He watched them on his feet. He watched them lying on his back. He watched them for as long as the star was visible in the sky, which was often. He didn't even realize Kylac and the raptor had gone until he tore his eyes away from the sun and tried to talk to them.

This time Stephen wasn't worried. Though he had seen enough *Scooby Doo* to know bad things always happened when the group split up, he trusted they knew what they were doing.

The dolphins swam along the surface. Stephen watched pods of several hundred form up, swim together, dive below the molten hot sea of hydrogen and helium, and then resurface. Individuals swam alone halfway across the surface, found a pod, and became lost in it. Some of them did flips and acrobatics in the air.

Stephen took to watching them without the binoculars. Now that he knew what to look for, he recognized them. The star wasn't just wavy and unstable—life was swimming around on it, and this planet orbited so close to its star he could see them with the naked eye.

Right now the human lay on his back, watching. He wasn't sure where they were going, or why, but there were millions of them, and they made the star itself appear to dance in the sky.

5

The rolling hills of plant matter had been kicked up by unimaginable winds centuries ago. The plants had retained that shape ever since. The Archeons lounged on one of these plateaus overlooking the purple landscape.

I wish I knew where they could have gone, Kuu gestured.

"I do not know either, but we will spread the word," said Bols.

"We're trying to pick up their scent again," Kylac said. "We don't even know where to begin searching."

If Friend really is trying to open a portal outside the universe... Her body halted. She could not finish.

"I saw the sphere he opened," Deka said. "It swallowed a star and half a planet. There was nothing on the other side. Even if it's not through time, it's dangerous."

"And Rive is half metal?" Bols said.

"Yes. Friend's first antisphere tore Rive apart. Friend had him rebuilt by a species on another planet. Then that world was destroyed, too. That's all we know about it."

"Friend claims he can't control it," Kylac said. "He can't stop meditating on it. When the equations begin returning their values and locking together, an antisphere opens. It destroys everything until he goes to a new planet, and then the equations begin again."

And yet you brought Stephen?

"He wanted to know what the universe really is," Kylac said. "We met him by accident. He felt so trapped by his limited experiences, and he was willing to dive into a larger ocean without any preparation. That kind of eagerness should be rewarded. We couldn't just leave him without answers. We can send him back to Earth as soon as things become dangerous."

"Could Friend be right?" said Bols. "Could he have found a way to make a portal beyond the universe and travel though time?"

"I hope not," Deka said, "because if he's right, then every Archeon in the universe is wrong about how the universe works."

Perhaps not wrong. Perhaps simply adding another layer to our understanding.

"Rel is still gone." Deka growled. "If I found how to travel through time, that would be the first thing I'd prevent."

Which is impossible, or you would not be here now.

"Which means time travel is impossible—and even if it's not, it's completely useless because you can't change anything!"

"We can't presume that," Bols said. "This is an area of study that has never been explored. If Friend is right, he may discover things work differently from what we have been taught."

I wish we could verify what Friend was doing. Learn where his antispheres lead, and why they are so dangerous and difficult to control.

Deka growled. "Thousands of people and at least four planets have been destroyed trying to find out."

"Why does he not settle on an uninhabited world and experiment there?" Bols said.

"He said he tried that," said Kylac, "but every habitable world has life on it, and it doesn't matter anyway because the ways he opens keep trying to pull the entire planet through, not just him. He'd quickly run out of uninhabited worlds to experiment on."

At least he and Rive evacuated Crexa first.

"If only he'd thought of doing that when he was on Rel!" Deka growled. "We've been to a hundred worlds. No survivors from Rel. It seems... He's doomed our race to extinction."

Everyone was silent for a few breaths, which, for the Hue, looked like dead stillness, not even breathing.

Kylac idly turned his muzzle to the star and watched the Faii. He picked out a few individuals and followed them as they swam across the surface and then vanished either into a pod or beneath the surface.

Kylac turned to Bols. He hadn't been with an Ori in many years, and he wondered if he could convince her to lie on him. He peeked from his sheath a little at the thought.

Suddenly the Heu began moving again. *I wish you had come here sooner. Something strange happened after the second disaster.*

Kylac's ears bloomed. Deka leaned forward.

Kuu danced away from them. *It resembles the antisphere you described.*

6

Stephen couldn't decide what was stranger: the people walking around without skin, or the dolphins swimming on the star above him. He made sure to keep his distance from the Heu as he walked by. He had left his backpack by the hub, confident nobody would take it.

He had been through several ways and seen different parts of the planet. What struck him most about this world was just how alike it was. The thick plants underfoot heaved into hills and dipped into valleys just like on Earth, but overall every part of the planet looked the same, and whether the sun was in the sky or not, the world was just as dim.

He walked by hundreds of lizards and skinless people. Many approached him to investigate, but only the lizards touched him, and it was always with the tongue. The Heu never touched each other, or the lizards.

All around him he saw groups of Heu dancing and groups of huge lizards munching on the ground they walked upon. They reminded Stephen of cows grazing and herons jumping about.

Nothing seemed to be happening. As far as the eye could see, nobody was doing anything but eating and dancing, and yet Stephen knew an entire civilization must be work around him.

7

Deka and Kylac stared into what appeared to be a fragment of an antisphere. As they walked around it, their minds struggled to comprehend what they were seeing. It was twice as tall as Deka and seemed ready to expand at any instant. It hovered just above the surface of the plants, a sphere broken into small lenses made of negative space-time—the outer shell of Friend's antisphere shattered into hair-thin sheets and suspended in mid-break.

Kuu and Bols stood in place as the Relians examined it. Kylac's fur had risen. Deka's killing claws were up.

"This has been here since the second disaster?" Kylac said.

"Correct," said Bols.

"This... This has not happened anywhere else."

"Has anyone tried to touch it?" Deka asked.

We have examined it, thrown pieces of plants into it. They disappeared.

"This is..." Deka began. He never finished.

The Relians stopped on opposite sides and gazed into it. They could see each other through the divides between the pieces of negative space, but the space itself was just void. It did not show them a view to somewhere else, but rather comprehension itself seemed to fall into these pieces.

"Is this what you saw?" Bols said.

"I'm not sure," said Kylac. "It could be related, though. The antisphere Friend opened pulled things into it, ripping them to pieces. None of that happened here, did it?"

"No. It just appeared here, and it hasn't moved since."

Kylac's tail slowly twitched from side to side as he thought. "Was a portal here?"

No.

Deka bent down and plucked a piece of plant from the ground. He carried it to the broken shadow of a sphere. From his point of view it appeared concave. Deka walked into it, holding the plant in front of him, and stepped very slowly. The tip of the plant disappeared into the shadow. Deka walked another claw's reach forward, then stepped back. The part of the planet that had entered the shadow was missing. Deka dropped the rest.

"Tried to pull our minds somewhere else..." Kylac said, slowly walking around it. "Pulled Rel somewhere else. Tried to pull our minds away, too. Deka... I think this is where it tried to pull our minds."

"Where?"

"Friend envisioned the universe in motion through some medium, going somewhere, and that is what causes time, so if he could open a portal outside reality, to a place where the universe used to be, he could travel backwards in time. What if this section of space has stopped moving?"

"But it is moving. It's here on an orbiting planet."

"It's moving in three dimensions, but not in time. When Genzin was lost, it was pulled somewhere else. Now there were portals to Genzin on Faii, correct?"

Kuu motioned *yes*.

"It could have pulled a piece of your spacetime with it, leaving this behind. A segment of space out of step with time."

All four of them stared at it for a few breaths.

"Deka..." Kylac continued. "What if this is a portal to another time? What if Friend is right? What if this *is* time travel?"

Deka huffed. "Looks useless."

"Uncontrolled. If you open a way to a random point in space, odds are it will be empty. Likewise, this could be an empty point in time as well. What if Friend is in the apprentice stage, still learning how to select a specific place and time to go?"

"But why here? This hasn't happened anywhere else."

"I don't know. We still don't have enough information."

All four regarded it. Kylac continued walking around it.

"I hope it closes on its own," Deka said at last. "Why is it still here? The last one closed on its own as soon as Friend was away from it."

"A piece of an unfinished equation trapped in time," Kylac said. "Maybe."

If Friend is an apprentice, what happens when he figures out how to be more specific? And how long?

49

Kylac's ears folded against his head. "We're used to calculating portals within spacetime. Friend is trying to separate the two. It should not be possible. What if it is?"

"It doesn't matter!" Deka snarled. "He's killing thousands of people! Discovery or not, if I smell him, I'm taking him down!"

"I agree," Kylac said. "Still... We both saw that antisphere. It went somewhere. Friend is on to something. Maybe it's time travel, maybe it's not. I wish we could find out without tearing entire planets apart."

Deka's voice was flat. "I could live my entire life without knowing."

What should we do about it?

Bols flicked her tongue. "Is it dangerous?"

"I don't think so," said Kylac. "It's not an antisphere, just the impression it left behind."

"It's very disturbing," Bols said. "Nobody wanted to be near it when it appeared."

"I don't blame them," said the raptor. "Something is definitely not right about it."

He thought about the antisphere on Crexa, how he stared at it as he stared at this remnant now. He thought about Rive.

"Kylac... Rive said Friend needed large planets that were far away from their star. Crexa fit that description. So did Genzin. But Rel did not."

Kylac had been circling the fragments and now stopped to face Deka through the broken, negative lenses. "Do you think Rive was trying to help us?"

"Why else would he say that?"

"That would narrow the possible planets they could go to a few dozen."

Deka curled his neck. "Better than six hundred and ten. If we go to those worlds first, maybe we can catch up to them before they destroy another culture."

"Good idea." Kylac's ears flicked.. "I know where we can start, and Stephen will enjoy it, too."

"The sooner I smell that fox, the sooner this will end."

Deka turned away from the shattered sphere and began walking to the portal, the others following. They were about sixty paces away when he heard Sonjaa's voice.

"Deka, help me!"

The raptor whirled around and landed in attack stance, facing the broken shell hovering over the ground.

"Deka, what's happening?!"

Deka saw her projected across the lens. She appeared vague and distorted, and when he looked directly at it, the image vanished, but he was sure it had been there.

"Friend! Rive! Help me, please! Someone! Help!"

Deka dug his claws into the plantsoil, jumped over Bols, moved to push Kuu out of the way, but she was swift on her feet and leaped aside just as Deka was on her. He bolted to the fragmented sphere.

"Sonjaa! I'm here! I'm coming!"

Deka had dashed about fifty paces—he was almost there—almost in reach of her voice—then he slammed into something, but nothing was in front of him. It was only empty space between him and the remnant, and yet he couldn't move. He reached out, clawed the air, stabbed the ground with his killing claws, trying to move forward, but he wasn't moving.

A crowd was blocking him. He smelled Relians all around. Deka felt he knew what was happening. People had gathered in front of him, keeping him from her. Sonjaa's voice screamed at Deka through the negative space. Deka screamed back, trying struggling to reach her with his voice as something held him back, and she was gone.

He felt something holding his tail. Deka turned his head and looked over his shoulder. Kylac was gripping Deka's tail with both hands. His paws had dug a trench thirty paces long into the plants, and he stood up to his calves in a pool of arsenic water.

Deka gasped, turned around all the way as Kylac let go of him. Deka grabbed his fox's arms and yanked him out of the pool. Bols instantly crawled to him and licked his fur. Kuu pranced up to him, opened her mouth for the first time, and licked his fur dry with both of her tongues.

Deka turned back to the remnants. They were dark and silent.

8

"She's alive," Deka whispered.

"There's nothing in there, Deka." Kylac moved his hand from his raptor's shoulder up his neck. "Those are pieces of empty time."

"You don't know that."

"We didn't hear anything."

Deka sat just a few paces away from the fragments of negative space, gazing between and into them. Bols and Kuu watched from a distance. The raptor had been there for two days, which wasn't very long, as a day on Faii only lasted a third of a day on Rel.

Deka shivered as he peered into the broken sphere, but he wasn't cold. Kylac could smell his fear. At first Kylac had been in favor of letting Deka follow through with this emotion, but now he began to question whether that was a good idea. Deka's fear made Kylac's fur stand on end.

"I saw her. She was in there. She was calling me. She even called Rive and Friend."

Kylac rubbed Deka's neck. It had always been a sure way to calm him down, but Deka was still shivering, his scent terrified.

"And for a few breaths," Deka continued, "she was here. She said my name. She's trying to reach me, too.

What if all I have to do is walk through it? What if she's on the other side? What if she's waiting for me? I'm staring into the abyss. I can't do it. I can't rush in, and I don't know why. Sonjaa is in there. She needs my help. All I have to do is walk through."

"Deka, stop thinking about it. Sonjaa is not in there. She's..." Kylac could not finish.

"She's what?" Deka stood up. "Since the first disaster you've been trying to convince me she's dead, but I heard her voice! I saw her in there! She is alive! She is out there! I won't abandon her!"

Kylac took a few deep breaths. "Wherever she is, it's not in there, and staring at it won't help you find her."

"You think she's dead—and what do you care? You don't have a mate! You don't care about anyone you've been with! You fuck 'em and leave 'em! You don't know what it's like to have a lifetime bond with anyone!"

Kylac reflected how infectious human vulgarity was. Deka smelled terrified, not angry, so Kylac did not flinch.

"You raised me to be that way, remember? And I owe you my life for that. I know exactly how it feels."

The theropod closed his eyes and looked away from Kylac, turning to the remnants. His knees gave way and he dropped to his side, still staring at the broken lenses. Deka began to make make quiet shrieking noises. They quickly rose in volume until they were so shrill and piercing the Faii probably heard him. Kylac held him as his scent oscillated between anger and grief.

"She's calling me!" Deka screamed. "I can't reach her! I can't—I can't—I can't get to her!"

Deka cried again. Kuu and Bols winced. It hurt their ears, and they only had a fraction of Kylac's hearing.

Kylac picked up Deka's ankles and began dragging him away. Kylac had pulled him sixty paces before Deka finally realized what he was doing.

"Let me go!" Deka shouted. "Take me back!"

Kylac kept walking, ears folded back, tail between his legs. Deka struggled, but he was weak, and Kylac held his ankles easily as the raptor screamed and reached out to the lens.

Eighty paces. Deka began to calm down. His cries now sounded more like a hatchling calling out for food than a raptor grieving the loss of a mate. Kuu and Bols now approached. Finally, after ninety paces, Deka fell silent.

Kylac let his legs drop, turned and stood over the raptor. Deka wasn't shaking anymore, and he smelled drained but not terrified. Bols and Kuu stood at Kylac's side, looking at Deka with intense concern.

"Feel better?" Kylac asked.

"Yes... Much..." He rolled over and raised his head, turning to each of them. "What happened to me? I thought I was over it. It's been so long... since I saw her on Xce. When I saw her die."

"Then you got around that thing, and it all came rushing back," Kylac said. "I figured it was doing something to you. I felt it, too."

What did you feel? gestured Kuu.

"Grief. Helplessness. That's why nobody wants to live near it." Kylac faced it now. From a distance, it did not seem so foreboding. "Maybe it's not an empty region of time. Maybe something's inside it. I don't understand what, but it may involve you, Deka."

Deka lay flat on his back, head still turned toward the broken sphere hanging in the air. His scent smelled defeated. "And Sonjaa. Kylac, thank you for pulling me away. Any longer and I think I would... I would've sat there forever. Or jumped in."

Kylac's tail waved.

9

Stephen sat on the arsenic plants, binoculars up to his face. He had come to realize he had a better view at sunrise and sunset, and on this world that happened every few hours with no difference in the light.

The stardolphins were swimming around, diving, jumping, flipping. Sometimes they covered the entire star's surface. Other times everyone was under the waves for several minutes, then the entire population rose to the surface and jumped at the same time. If Stephen were paranoid, he would have sworn they knew they were being watched and were putting on a show just for him.

He wondered if any of them were watching this planet. He wondered if they were just as baffled by the Heu and Ori as he was. From time to time he Stephen turned away from the star and observed the natives of this plant world from time to time. He saw movements he did not understand, actions he did not comprehend. He turned to the star again, and he saw more things he did not understand on a place he could not imagine living.

Stephen sat here for days. Deka and Kylac had told him what to do when nature called, and his time in the Army had taught him how to live without the comforts of home, so he wasn't concerned about it. He was starting to feel dirty already, and it was just like being in the field.

Stephen was thirsty, but not dying of thirst yet. The water contained arsenic, so he dared not drink any. He trusted Deka and Kylac had planned for that. The Selts didn't take him apart just to measure vitals—they knew everything about him, and so did his companions. They knew what he could drink, what meat he could eat uncooked, which fruits and vegetables he could have, and which he should stay away from, so Stephen didn't eat or drink without asking.

He smelled dog sitting next to him, and he felt the air move out of the way for a large reptile lying down on his other side. Stephen lowered the binoculars and turned to each of them.

"Hi, guys."

Deka looked like death.

"What happened to you?" Stephen asked.

"I don't know." Deka's voice sounded distant and fatigued.

"Have you been watching the Faii this whole time?" Kylac said.

"Jus— How is this possible?!" Stephen flailed his limbs as he shouted. "That's a goddamned star! Nuclear fusion! How the hell are they alive? How did they get there? What do they eat, and why do they look like dolphins?"

Kylac waved his tail. Deka held his claws together, his whole body sagging as if he didn't have the energy to laugh anymore.

"Life is full of surprises," said the fox.

"Is that all? We can't go there, send a message, or something?"

"It's a star, Stephen," Kylac said. "A cold star, but still hot enough to boil our hides. Archeons have tried to reach out to them, but nobody has received anything we can call a response."

"Do they know anything about them?" Stephen said, gesturing to the skinless bipeds and the monitor lizards wandering around.

"Very little," said Deka. "We know they exist. It seems impossible, and yet they live."

"I can't stop looking at them! This has to be the most incredible thing I've ever seen, and I don't even know what I'm looking at!"

"That will happen frequently at first," said Deka. "Gradually you will learn how to perceive the world from other people's point of view, and things will not be so mysterious anymore."

Stephen offered Deka the binoculars. Deka took them, held one of the eyepieces up to his eye, and turned his muzzle upwards. He flinched, handed them back to Stephen.

"I can see them just fine."

Stephen passed them to Kylac. He refused them at first. Stephen smiled. "Come on, fox, see the world as I see it."

Kylac's tail waved a little. He took the binoculars and looked at the star.

"Fuck!"

Stephen laughed. Hearing them use English curse words was even funnier than seeing them watch TV.

Kylac handed them back to the human. "I don't know how you live with those eyes, and they used to be worse."

Stephen was still laughing. So was Kylac. Deka held his claws together. All three of them watched the setting star as the Faii made their hypnotic patterns on the dim surface.

Nule

The sphere opened on a bright world. Stephen stepped out of it. Deka and Kylac emerged after him, and the portal closed.

"So normally there would be hundreds of spheres on each world, all leading to different planets, and people could just walk to any world they wanted?"

"That's right," Deka answered. "You joined us at a very bad time for the contacted universe. Everyone is afraid of making ways to other worlds since the disasters."

"And now the story we're telling only makes them more nervous," said the fox. "Everyone is isolated now. That's why we have to take you around like this. Just a couple years ago, you would've been able to jump to a hundred worlds in as little as an hour."

"Wow. And that sun is blinding."

The Relians scented the hub area, Deka sniffing the air, and Kylac scenting the ground around the paths where the portals used to be. Stephen watched them for a moment.

"I don't smell Rive or Friend," said the fox.

"No, they weren't here."

"Would they have come here?" asked the human.

"We're following a lead," said Kylac. "We thought this was a place they might have come."

"And while you're here, you can meet a noncompatible species, Stephen," said the raptor.

"Don't you mean incompatible?"

"That word has technology connotations. When referring to cross-species, the correct word is noncompatible."

"Sexually," continued the fox. "Remember I told you about that? The sexual system forms the foundation for how a species understands reality. Now you can meet a species that is not compatible with us."

They began walking, the Relians flanking the human.

"Well, this place seems normal enough," Stephen said. "Bright sunlight, soil, plants, trees. Gravity's is a bit strong though. Feel about twenty pounds heavier. Who lives here?"

"You'll see," said Deka.

"As long as everybody has skin, I think I'm ready. So when the portals were up and everyone could come and go as they pleased, weren't those people afraid someone will come in and invade them?"

"That is a common theme in your movies," said Deka. "You yearn for contact, and yet you fear it. Why do you assume an alien civilization would want to take over your planet and enslave or exterminate you? No species has a reason to do that to anyone."

Kylac knew where Deka was going with this and picked up the thought. "Historically, you have taken over other nations, enslaved and exterminated other humans, so naturally you fear this will be done to you. It's more fear of the worst of your own nature than anyone else's."

"How do you know that?" Stephen said. "Did you read up on our history, too?"

"Actually no," Deka said. "We know little of your kind's history, but we can infer it in a general sense because it is developing on a path common among all who live without a companion race."

"We're predictable?"

"Yes, you are. The contacted universe has witnessed many species such as yours, and all of them develop in similar ways. There are major differences, but broadly, you have the same story."

Stephen was silent for a few paces. Kylac took the time to speak up.

"The people you're going to meet now will definitely surprise you. Probably as far from what you expect as you can get and still be up close to them. These people are so alien they have no language we can translate."

"You can't even speak their language?"

"Their language isn't spoken," said Kylac, "and it's not body language."

They did not have to walk very far from the hub before they encountered transparent blobs all over the ground, their organs plainly visible and never arranged in the same way between two individuals. Stephen knelt beside one of the blobs and looked it over. Kylac and Deka joined him. The thing was twice as large as Stephen's head, and only three claw's reaches high.

"What is this?" Stephen said.

"This is a Type One Nule," Kylac answered.

"Is this a single-celled... something?"

"Not quite. They are multi-celled, but in their case, becoming a complex organism meant evolving into a larger version of a single cell."

It twitched and extended a pseudopod and began flowing away.

"It's a blob," Stephen said. "It's intelligent?"

"Enough to make portals across the universe, yes," said Deka.

"Where is its brain?"

"Their nervous system," said the fox, "is spread out across their entire body. Watch this."

Their blob was moving toward another. Moments later, they met. Their cell membranes joined, and they began flowing into each other.

"What are they doing?"

"The Type One Nules speak by trading memories. Those are contained inside their cytoplasm. Do you know what that is?"

"I remember. Hasn't been that long since high school. But... They trade memories?"

"If you think about it, Stephen," the fox said, "it's what we do when we talk. They're just more direct."

"But I don't give away my memories when I talk to someone."

"Oh, they give everything back. It's not permanent."

They knelt and watched the conversation. Cell matter flowed between them, circulated around, and then flowed back into the individuals.

"So can we talk to them at all?" Stephen said.

"A few species can, to an extent," said Deka, "but very little they say to each other can be translated. We can translate a few concepts, but beyond that, no. They experience reality much differently from how we do."

"And they reproduce asexually," Kylac continued. "When they're ready, they just divide—but here's the interesting part. Each Type One decides which traits to pass on and which to leave off."

"They can decide what kind of children they'll have?"

"They have absolute control over their bodies. Which memories to trade, which genes to copy, food to process, and everything else. When microbes invade them, they control how their immune system responds."

"Try to imagine what that's like," Deka resumed while his fox took a breath. "Having that much control of your body. Being aware every time your stomach secretes digestive chemicals, and how much. Aware of how much blood your body produces. How progress is going repairing that bruise. It's an incredible idea."

The two organisms finished trading bodily matter, and they parted ways smoothly.

"Do we know anything about them?"

"Almost nothing," said the fox. "Believe it or not, life is common in the universe, but quite often we can't understand it. They know we exist. We know they exist. Beyond that, all we can do is look at each other and make gestures of surrender."

"Shrug?" Stephen demonstrated.

"That's it."

"Does this one even know I'm here? Where are its eyes? Does it have ears, or a nose, or a mouth?"

"They do not use those senses, Stephen," said Deka. "In fact, they have only two. Sensing each other, and sensing the outside. Go ahead, touch it."

"Can I?"

"Sure, you won't hurt it."

Stephen extended a single finger, reached down, and pushed. The body reacted by opening its cell membrane, pulling his finger into its cytoplasm, and moving fluid around it. After a few breaths, it restored the barrier and excluded Stephen's finger. He raised it to his eyes, turned it around. It wasn't even wet.

"What just happened?"

"It sensed you," Deka said. "It knows you're here and not of its kind."

"That's it? I just made contact with an in— A noncompatible species?"

"Yup." Deka clicked his claws.

"I was hoping it would do more than that. Like, grab my arm and start digesting it or something."

"It has no way to communicate with you, Stephen. It's probably just as frustrated as you are. All it can do is sit there and know you exist, but it doesn't understand what you are. Their way of knowing the universe is so different from ours that ideas cannot be shared or understood."

"But they make portals?"

"They do indeed." Deka rose. "Funny isn't it? Sometimes learning how to sense the true structure of the universe is easier than understanding the different creatures that inhabit it."

Kylac rose to full height. "Now we'll take you to meet a Type Two Nule."

Stephen stood and followed the Relians through the field of blobs, stepping over them carefully. He was wearing clothes and shoes, so he was much less sure of where his feet were.

They walked over and around hundreds of blobs, and then finally Kylac stopped over one was covered in fleshy, grey skin. It didn't flow, but rolled from place to place. Deka knelt over it beside Kylac, and Stephen knelt on the other side.

"Type Two Nule," Kylac began. "These people did develop a protective layer of skin. They don't share thoughts directly. Any guesses how they communicate?"

Stephen thought about it. Then he looked at Kylac. "Sharing brainwaves?"

Kylac waved his tail. "If only. No, they leave a trail of scent on the ground as they move. Others pick up on the trail and sense it."

"That's kinda gross."

"It gets better," said Deka. "They reproduce sexually. This one is male. Can you tell?"

The skin was covered in hundreds of tiny appendages. Stephen looked it over from one side to the other.

"It's covered in dicks?"

"Right!" Kylac's tail whipped from side to side. "That's exactly right! You can tell a female because it's covered in vaginas. Males roll into females blindly, and during the mating season, they release sperm from every appendage at once. Odds are some will make it into the right place."

"That's also gross."

"You can touch him, too, if you want."

Stephen looked down at it, then up at Kylac.

"Oh, come on, Stephen, you touched mine. How is this different?"

"This is just weird."

"Life is weird," Deka said. "We don't know exactly how these two came together and realized the other was sentient, but somehow they figured out how to communicate. Learning to understand such different perspectives on the universe pushed them to understand how the universe really is, and they discovered portal physics."

"So what did they do when they left this planet? How did others react?"

"That was long before our time," said Kylac. "I've learned it was very frustrating. Alien life, and we can't understand it. I imagine it's just as frustrating for them. Fortunately, the contacted planets did know of another race they could relate to back then."

"There are others like this?"

"Oh yes," Deka said. "Several worlds have these kinds of life forms, and they can communicate with each other. Not always easily, but they're not alone. This is fairly normal. Like species tend to group together. There's a group for the gas giant species, a group for the oxygen breathers, another for methane breathers, another for ocean dwellers. We do everything we can to learn about one another, but sometimes all we can do is stare at life on other worlds and... shrug."

Stephen gazed at it for some time, watching it undulate. He extended a finger and pressed the Type Two lightly on the surface of its skin. It vibrated. Liquid oozed out of its many appendages. Stephen withdrew and held his finger up to his nose. It smelled like boiled cabbage and stale urine.

Deka and Kylac were rolling on the ground, Deka scraping Kylac's arm with his hand-claws while clicking his toe claws together, Kylac waving his tail so fast he might have taken flight.

Eiae

I

Stephen landed in total darkness. He heard Deka and Kylac land next to him, but he couldn't see them because it wasn't just dark. It was completely, absolutely black.

He heard his companions sniffing around the area, and Stephen held perfectly still, waiting for them to figure out if Rive and Friend had been here or not. After five minutes of sniffing noises, Stephen felt the need to speak.

"When does the sun come up?"

"It's already up," Kylac said.

"Where?"

Stephen felt a three-fingered hand on his shoulder. "We call this world Eiae. Don't be afraid. We'll guide you."

Deka started leading him forward. Stephen trusted his companions, but every blind step tested his faith. He felt like he was perpetually on the edge of a huge cliff and he would fall off if he took one more step. He frequently stumbled to a halt and felt the air in front of himself, Deka nudging him on. After a breath, Stephen felt safe taking a few more steps, each one like walking into a solid black wall, leaving him surprised when he did not collide with anything.

"Life requires extremely specific conditions to develop and survive," Deka said. "Not always the same conditions though. Many species consider what you call visible light highly radioactive and dangerous. Some of these species see by x-rays. Their planet is close to its star, and their atmosphere developed in such a way that blocks light but allows x-rays through, and their kind has adapted to use those as a sense. It's difficult to know how they understand their world because we have nothing to compare it with, but they find us just as fascinating."

A padded hand touched Stephen at his back, and now Kylac spoke.

"Transplant them to a planet with light on it, and they will die. Place humans on their world, without your specific conditions for life, and you will meet the same fate. We are all products of our environment."

"Is that where we are? Guys, you know I can't live on a planet with lots of x-rays."

"We know," Deka said, claws poking Stephen's shoulder. "We wanted you to know such places exist, and life has evolved means to survive there. Take a moment to think, Stephen. What's missing from this planet, besides light?"

Stephen couldn't think of anything besides the blackness he walked into. This wasn't just nighttime darkness. It was the kind of dark where night vision goggles would not help. It was the kind of dark that made you question the existence of your own limbs. His footsteps made sound, but there was no sense of movement, and darkness became a solid that entombed him.

"It's very…" he swallowed. "This is terrifying, but… sounds are clean. Everything seems so loud here. I'm scared out of my mind, but it's kind of relaxing, too."

"That's because this planet is completely free of electromagnetism," Kylac said.

"Well, every planet's been free of electricity," Stephen said.

"Not just electricity," Kylac said. "All electromagnetic activity. Eiae's atmosphere keeps out all energy from its star, and from the rest of the universe. It does not even have a magnetic field."

"You don't notice the enormous amount of energy around you on Earth," said Deka. "It's everywhere, and you use it for everything. Television, radio, satellites. All of this energy is ricocheting around and through you day and night. Even Earth's magnetic field affects you. You assume all of it passes straight through you, but on a subconscious level, you are aware of it, and it does affect your society."

"You are about to meet two species who can never live on Earth because they are sensitive to electromagnetic radiation. It actually makes them sick, and they can die from overexposure. They consider it poisonous."

"So what planets do they visit?"

"They have their own group of worlds they travel to. Some are brave enough to condition themselves to visit worlds with atmospheres that allow radiation in."

"So... So how the hell are you two getting around? There's not even enough light for night vision, is there?"

"Scent still exists here," Deka said. "We can navigate with sound, too, but not nearly as well as these people."

They stopped. Stephen stood still. As far as his senses could tell, he hadn't moved an inch.

"Say hello to the Heeke, Stephen."

"Hello."

Shrill screeches and clicks answered him. They weren't loud, but somehow they were piercing. Noises rose up from the ground and took to the air. Stephen felt like their noises were focused on him. They moved about him like a carousel, the sound pushing Stephen down, and it became a struggle to stand straight. Only Kylac and Deka held him upright.

"What's happening? What are they saying?"

"They haven't said anything yet," Kylac answered. "Those are sonar clicks."

The sound of flapping wings felt familiar. They were all around him, buzzing his head, flying off, circling back, piercing him with more sonar.

"Are they bats?"

"They are similar," said Deka, "but bigger."

"Jesus... What are they doing?"

"They're looking at you. They've never seen anything like you before, and they're curious."

"They also smell your fear," said Kylac. "They can tell you're from a planet with light."

"How?"

"Humans can't see in the dark. You have a deep-rooted fear of the dark because you are defenseless inside it. Take a moment to think how they know the world. Sound waves are outlines to them. Scent is color."

"They have no eyes," Deka said. "No eyes at all. They don't know what sight is. Their word for sight describes what the Wen do to hunt, not as you understand it."

The whole time they were talking, the flapping and swooping and piercing clicks and whistles bombarded Stephen. He flinched constantly, standing upright in Deka and Kylac's grip.

Gradually the Heeke landed, and Stephen heard shuffling across the soil. He felt dozens of noses close to him, taking in his scent. Stephen heard grunts and clicks.

"Now they're speaking," Deka said.

Grunting and clicking and long strings of syllables that sounded like Hawaiian played at triple-speed.

"Guys, I don't like this."

Sounds came out of Kylac that Stephen did not think he was capable of making. Stephen turned to him but wasn't sure if he was facing the fox or not. He was starting to doubt Kylac was even there.

Deka also made the noises.

Hundreds of voices replied. Stephen wanted to cover his ears, but he knew it would not help—the sounds didn't need to enter his ears to be heard.

Deka picked up one of Stephen's hands and held it out for him. A second later, something warm with prickly hair pressed against his palm. Stephen gasped and pulled away, but Deka extended his hand again.

Stephen wasn't sure what he was feeling at first, but Deka guided his hand across the fur. Whoever this was stood very still, and gradually familiar anatomy began to fall into place. He recognized shoulders. A neck. Large pectoral muscles. Deka guided Stephen's hand down the arms, which ended in fingers, but the fingers formed the tips of fleshy wings. They had no hands. This one felt vaguely humanoid—probably the most humanlike species Stephen had encountered since he left Earth.

Deka guided his hand over parts he didn't think he should feel. This individual was female. She didn't seem to mind Stephen feeling her there, but Stephen still felt weird.

"That is a Heeke," Deka said.

"You're right. Kinda like a bat. But she feels human."

"She does resemble one from the waist up, yes. They find it fascinating that you can't see them, since they can see you just fine."

Deka lowered Stephen's hand and made some noises to her. Stephen heard her feet moving away. Everyone was sending sound waves at Stephen, though not as shrill as before.

"Next you'll meet the Wen," Kylac said as they began walking him again. "But first, feel this."

Kylac now took Stephen's hand and placed it on something hard and vertical. He guided his hand up. Up. Up. It was solid, and yet it moved back and forth.

"What's this?"

"This is how the plants survive. Eiae is covered in plants, but they need light. They collect it by launching tendrils high into the atmosphere. This thing you're feeling is one of those. Much of the planet is a thin forest of these streamers. They collect light and channel the energy down to the plants on the surface. Supports the whole ecosystem. The Heeke feed on the fruit. The Wen once fed on the Heeke."

"They did? They don't hunt them now, do they?"

"Not since they realized the Heeke were intelligent."

"That's... Wow. And the Heeke aren't afraid of them?"

"They found other things to eat."

"So what do the Wen look like?"

"We don't know." Stephen turned to face Deka's voice. "The Wen have never ventured onto a planet with light. Nobody has seen them, and they do not like to be touched. The only ones who can touch them are the Heeke."

"Will I meet one?"

"You already have. Capca is right behind you."

Stephen turned around, but he only saw a wall of black. He heard nothing. Felt nothing. He couldn't even tell which way was up anymore.

"Hello, Stephen." The voice sounded masculine and large. Now Stephen felt a looming presence before him.

"Hi..."

"I am Capca. One of the Archeons for this planet."

"You learned English quick."

"She's an Archeon," Kylac said.

"She?"

"I have been listening to you speak since you arrived. Your language is similar to a few others I know, so it is not difficult to learn."

Stephen was trying to breathe regularly. "I can't tell... How large are you, compared to me?"

"I stand about twice as tall as you. I have six legs. I do not know how to describe myself in terms you will understand. I do have eyes, though."

"Really? On a planet without light?"

"This world once did have light. Gradually it faded until there was nothing left. The Wen have been around all that time. We can still see."

Stephen felt a reptile muzzle by his ear. "Remember Stephen, nothing here will hurt you. Try to see the world as they see it. Do not let primitive fear override your higher mind, and do not use that flashlight you brought. You will hurt them."

A fox muzzle in his other ear. "Capca will guide you now. We'll be back for you soon."

The hands fell away from his back. Stephen floated in a void now, and he reached out and tried to grab something. "Wait, guys! Wait! Oh God, please don't leave me alone now!"

Stephen heard footsteps trotting away, probably back to the hub. He stood perfectly still.

"I'm told this is your first time off your planet," said Capca. "You're an isolated species. Why did Deka and Kylac take you with them?"

Stephen made a conscious effort to keep his breathing normal. He sat down, which helped remind him he was not in a void and he did still exist.

"I begged them to take me with them. I had to know what things were like out there. All I've ever known is my little country on my planet."

"So you are probably the only one of your kind I will ever meet. Interesting. You resemble a Heeke without wings or fur."

Stephen felt movement around him. This person was massive, whatever she was. The voice came from far above his head. Her footsteps felt and sounded heavy, but she was always too far away to reach out and touch.

"They told me you used to be a predator."

The voice circled Stephen. "I still am. We just don't hunt the Heeke anymore. Haven't in many generations."

"Um... That's making me uncomfortable. I'm not used to being alone with a predator."

"You seem fine with Deka and Kylac. Both of them are predators."

"Not Kylac," Stephen snickered. "He's a lover."

"Oh, Kylac is a predator, too."

"He is?"

"You didn't know?"

"Well, I kinda figured... But I've never seen him hunt anything. Deka always does that."

"I should not say more on the subject. It's not my place. To answer your question from before, yes, I can see. It is actually how I hunt. I project electromagnetic waves out to stun prey, and then I kill."

"Right. Listen, please don't be offended. I ask because I don't know. How does prey make friends with their predator? If it was me, I'd never trust anything that can kill me."

Her voice became stationary now,, from high above him. "It is what happens when two species want to learn more about each other. We overcame our base instinct."

"It doesn't seem like it could be that easy."

"Follow me," she said.

"Fo— What? How?"

"Use my voice as a guide."

"But... But I..."

Stephen did indeed hear her footsteps walking away from him. The thought of trying to walk in total darkness made his heart skip beats, and his lungs did not seem to take in any air at all.

"There is nothing between us. Walk toward me."

Stephen stood, held his hand out, and took a step. He braced himself to collide with something, and when he didn't, he prepared to fall off a ledge. Neither happened, so he did not know what to do.

"Don't be afraid, Stephen. You're not the first visual species I've met who had a difficult time learning how to use other senses. Deka and Kylac brought you here to experience life off your planet."

Stephen took a few deep breaths and walked toward her voice. As Stephen neared, she moved away.

"I will tell you a story on the way to help you navigate," said Capca. "Listen closely."

Stephen adjusted course until her voice sounded centered.

"It's the story of how the Heeke and the Wen came together."

2

Deka sat on the edge of a lake. He could only tell it was there by the smell of the algae in the water and the sound of the lapping against the shore. It was refreshing to shut off the eyes for a while and let the other senses strengthen.

Beside Deka, in the lower branches of one of the streamer trees, hung Kylac and Lalks. The Heeke Archeon had been rather fond of the canine's pheromones since meeting him twenty minutes ago, and now he hung upside down, wings wrapped around Kylac so tight the fox could not move at all, penis buried as far into Kylac as it could be.

Ollet, the other Wen Archeon, lay next to Deka. Ollet smelled amused listening to the fox and the Heeke. Deka rubbed his claws. It was cute in a way, Kylac hanging upside down, a Heeke's wings the only thing keeping him from falling.

"Qan asked about you," Ollet said, her voice strained with age.

"Qan?" Deka said. "She came here from Neben?"

"She's searching for you two. Said you needed to return to Neben as soon as you could. They found something new."

"Did she say what?"

"No. Or rather, she wanted to, but our language does not have words for what she wanted to tell you. You should make that your next destination."

"Neben," Deka sighed. "Rive and Friend wouldn't have gone there, but we'll make that our next stop."

"Do you plan to take the human?"

"We are actually searching for Rive and Friend. We're just taking Stephen along for a few planets until we find them. He wanted to know what life was like off his world. Have you heard anything of Rive or Friend?"

"They have not been here, and we have had no unusual spheres on this world. From the story you tell, he sounds dangerous. Possibly insane."

"He believes he's on to something," Deka said, growling, "and he doesn't care who he hurts as he figures it out. Rive may be the only one keeping him in line, since he was the one who evacuated Crexa."

"I was afraid to make any ways to other worlds after hearing three planets couldn't be reached anymore. I wondered what happened. The answer is more disturbing than anything I imagined."

"I'm sorry we didn't check in on you," Deka said. "We figured everyone here would be fine."

"We were, although Capca was unconscious for so long we feared she would never wake."

"Has anyone had difficulty keeping offworld portals open?"

"No one has tried."

"Kylac and I... We seem to have lost our ability to keep ways open long-term."

"I am sorry. I dislike being confined to my own planet. Everyone has been restless since this began."

In the tree, Lalks was making louder noises, and Kylac was moaning with him, barely able to breathe. Being wrapped up tight, upside down and helpless in Heeke wings was apparently a huge turn-on for him. Lalks was finishing, but the fox hadn't even started.

3

Taking blind steps into an abyss never became comfortable. After the first fifty paces, he became confident, and then he collided with something, which made him even more afraid than before.

He bumped into Heeke constantly while feeling his way, their noses and wings brushing against him. He was glad he didn't have to go naked here, but even fully clothed he felt exposed and helpless.

He wanted to use his flashlight so badly. It was right there on the multi-tool in the pouch attached to his belt. All he had to do was reach down and take it out. Everything would be easy if he could only do that, but Stephen resisted. Deka told him nobody would hurt him, so he hoped that meant someone would rescue him if he really was about to fall off the edge of a cliff.

Capca walked ahead of him, her voice guiding him in a general direction. She told him of the early days of civilization on this planet. The Wen had their society, the Heeke had their own, and they intersected only with the Wen hunting and killing the Heeke.

The Heeke lived in the air, often making homes in the thicker streamer stalks that reached high into the atmosphere. The Wen lived on the ground, waiting for one of the Heeke to come close for a bite of fruit.

A Heeke named Uvir was down in the branches of one of the trees, eating the fruit. She was young, but still had enough sense to stay above the ground. Hunting her was a Wen named Tels.

Tels sat and watched the Heeke for a while but did not project an energy wave at Uvir yet. She was the first to sit and watch the Heeke instead of hunt them. As she watched, more Heeke descended and began eating the fruit. Tels began to observe that many of the things they did were not animal, which made her curious.

So she approached the Heeke in the tree. She made herself known, but she did not attack. She watched. She listened. She began to hear some sounds more than others, which implied language.

Most of the Heeke flew away, but Uvir remained. She was doing something the Heeke had never done before: observe their predators instead of retreating to safety.

Wen and Heeke watched one another. Finally, Uvir spoke to Tels. She picked up a piece of fruit, said her word for it, and then tossed it to Tels. The predator picked it up in her mouth. It was poison to her species, so she did not eat it, but she thought she understood, and said her word for the fruit.

Civilization began in that moment. Hunting gradually stopped. Fear quickly gave way to understanding. They were so different, but so much alike as well. The Heeke did not want to be afraid of the Wen, and the Wen did not want to destroy the Heeke. The Wen learned how to perceive the planet from the sightless point of view of a Heeke, and the Heeke learned of a sense they themselves did not possess.

"This level of understanding took generations," Capca concluded. "Once we had it, we learned how to understand the universe in the same way. We discovered portal physics, and we began exploring the perspectives of species all over the contacted universe."

Capca's voice held still by the time she reached the end of the story. Stephen had navigated around dozens of trees, small plants, and people. He had bumped into most of them along the way, but he was unhurt and slightly less afraid of the abyss.

"There is a plant in front of you," Capca said. "Touch it."

Stephen stepped forward, arm outstretched, and felt in front of him. His palm touched a smooth surface. It reminded him of a tree.

"Your kind appears to have evolved from tree-dwellers," she said. "Climb it."

"Uh..." Stephen muttered as he felt around. He found a low branch, hoisted himself up. It had been a long time since he'd done PT, either in daylight or nighttime, so it was not as easy as he thought it should be. The branch was low enough and sturdy. Thin, and yet it did not seem to give. He sat down on it, felt around with a free hand, finding more branches nearby.

Capca's voice came from Stephen's eye level. "You are in the tree I told you about, where Uvir first spoke to Tels." She sounded like she was an arm's reach away, and he was tempted to reach out and try to touch her, but he held the branch instead.

"This is the same tree? It's been alive that long?"

"Certainly."

Stephen was about to express wonder at that, but then he remembered there were redwoods in California that had been alive when Jesus walked the Earth. Without the need to grow taller and taller in competition for more sunlight, trees could survive even longer.

"You are sitting in the birthplace of our civilization. This tree became the neutral ground between our two species. You cannot see it, but it is the only one nearby. It was here, in and around this very tree, that our languages were translated and we began to learn about one another."

He ran his hand down the branch. It felt smooth, like birch, but not rigid. "It's unbelievable. Where I come from, predators don't just stop hunting their prey. Prey doesn't just stop being afraid of getting eaten."

"Try to imagine it. The creatures you have feared your entire life. What would happen if you realized they could speak?"

Stephen sat there. He felt her breath. She was right there—she could be just inches away from his body, and he couldn't see her.

"I'd still be terrified. How do I know they won't attack me while I'm trying to learn about them? And where I come from, most predators can see in the dark. We can't. It wouldn't be very fair. There'd have to be a lot of trust on our part, but none on theirs."

She made a sound from the back of her throat. Stephen got the feeling she was laughing. "It really is beyond your comprehension, isn't it? I confess, you are the first uncontacted species I've ever met. You are said to be paranoid because your kind is constantly hunting itself. Even herbivorous species will do so if left on their own. You cannot even hold it in your mind that something which can harm you will choose not to."

Stephen did not know if that was a question or not. She remained silent for a moment. So did he.

"Why are you here?" she said.

"To find out what I'm missing."

"Are you afraid of me?"

"Yes."

"Even though I have not harmed you, have guided you to a safe place, have shown nothing but kindness, you still fear me?"

"You, the darkness, everything. I've never been more scared in my life."

She did not answer. Stephen hoped he hadn't said the wrong thing. After a minute, he began to wonder if she was still there.

"Tell me a story of your people, Stephen."

"Sure. What story?"

"The greatest story you know."

Stephen thought about it for a moment. "We have a lot of stories. I'm trying to think of something you'd understand."

"Does not matter if I understand. I want to hear it."

Stephen slowly lowered himself from the branch to the ground. He remembered the climb had not been very far, but because he couldn't see, his senses told him he was about to fall into a bottomless pit.

He let go and dropped two feet the ground. He straightened up, proud he managed to do that with dignity. He felt Capca moving far above him now, still looming. He could think of only one story worth telling.

"Old Marley was as dead as a door-nail. You must understand this, or nothing wonderful will come of the story I am going to tell."

4

Kylac was now on the ground with Lalks underneath him, holding his wings down and his legs out.

"Should we help him?" Ollet said. "Lalks seems rather distressed."

Deka replied in English. "No, Kylac's just fucking his brains out."

Low, throaty sounds came from Ollet. She replied in her native words. "Stephen's language is very blunt, yet figurative."

"I know!" Kylac said. "It's so full of metaphor and imagery!"

Lalks moaned and tried to wrap Kylac up in his wings, but Kylac was holding his arms down while he thrust. Mating flat on the ground was an exotic feeling for the Heeke.

"They're a sight-based species," Deka said. "Some of the other languages on their planet are just as expressive. We only had time to sample a few of them."

"They're very difficult to understand," said Ollet. "They rely so much on light and not enough on sound. You said you left him with Capca?"

Lalks was making more noise than seemed natural.

"Kylac," Ollet said. "I think you're hurting him."

"I'm giving him the best he's ever had!"

Kylac did something with his hips. The Heeke made a strange noise that lasted for many breaths. Kylac grunted a few more times.

"I... didn't know the Heeke could make that sound," Ollet said.

Deka didn't know a Heeke could make that sound either. They waited until the fox was done before discussing anything further. Finally Kylac withdrew and stood up. Lalks lay on the ground for a few breaths and then sat up.

"Are you all right?" Ollet said, moving over him, scenting him from ears to feet.

"I'm fine," said Lalks, voice dreamy and wavering. "Never been with one of the Relian canines before. I had to try."

Ollet made sounds from her throat. She directed her voice at Deka. "To be young again."

Deka clicked his claws. Kylac's tail thumped the ground. The Heeke merely sat still and swayed back and forth, reeling from strange climax.

"This world matches the kind of place Friend is likely to go," Deka said. "If you smell him or Rive, don't hesitate to kill them."

"Kill?" Ollet said.

"He's that dangerous," Kylac said, mouth still agape and panting. "He's destroyed at least four planets. We're not sure where he's going, but we think we've narrowed it down."

"I wish I could help," Ollet said.

"You can help by going to a few worlds and spreading the word," Deka said.

"I'll go," said Lalks.

"You will not," Ollet snapped. "You aren't ready."

"I've been getting ready for years. I've been asking the Wen to see me for longer and longer. Ten of them at once saw me, and I didn't even flinch. I can handle a planet."

Deka and Kylac sensed an electromagnetic hum coming from Ollet aimed at Lalks. The Heeke was indeed not flinching as the Wen saw him.

"Didn't even feel sick!" Lalks said.

Kylac faced him, using his nose to guide him. "Lalks... You've never been to a lighted world?"

Lalks made a sulking noise. No body language. Their gestures existed purely in the auditory realm. "I want to, but they won't let me!"

"For good reason," Ollet said, lying down next to the Heeke. "Even ten of us at once can't produce a fraction of the energy you will endure on a planet with a transparent atmosphere."

"I'm an Archeon now! It's been years, and you still won't tell me how to get to a lighted world!"

"You are not even ready to prepare for one."

"When will I be ready?"

Ollet did not answer right away, and her scent became something almost maternal. She shifted, pulled Lalks close. Deka soundlessly held his claws together. Kylac tried to keep his tail still.

"Lighted worlds are not safe for any of us," she said. "There are other worlds you'll have to visit first. Easier worlds for our bodies to handle."

"That won't help Deka and Kylac. I need to go now!"

"You don't know what you're asking for," said Ollet, her voice coming from Lalks' height. She was rubbing her cheek against his. "A lighted world will kill you."

"Most Heeke can't even stand up when one Wen sees them. I can do ten."

"Yes, and imagine what an entire planet bathed in it will feel like."

"I can handle it!" Lalks faced the Relians and did not make a sound. Deka and Kylac thought they understood what he was asking.

Kylac elbowed Deka in the ribs. Deka lay his claws on Kylac's shoulder.

"We'll give you a taste of a lighted world," said the raptor.

"You will?"

Kylac continued the thought. "If you can tolerate it, you're ready."

Ollet hissed, a gesture of shock and anger. Deka audibly rubbed his claws, and Kylac let his tail thump the ground. Ollet's hissing subsided.

5

Stephen had intended to tell a quick version of the story, but as he progressed, he found himself acting it out. He had seen the movies so many times he knew the story line for line, so he became a one-man play of *A Christmas Carol*. Stephen didn't know he had it in him, but he himself became swept up in the story, and he started believing in the dialogue and the narration and even the gestures he performed as he hopped around, switching characters and voices. Something about the knowledge that this might be the only human story Capca would ever hear drove him to tell it as well as he possibly could.

Finally, he reached the end. "And so, as Tiny Tim observed, God bless Us, Every One!" He stood in silence, beaming with pride and wondering why he had never thought to try acting as a career.

"What is a spirit?" Capca asked.

Exhausted, Stephen sat down on top of his backpack, facing her voice. "A ghost. It's an old belief that the dead exist as spirits. They're alive without their bodies, and they can harass the living."

"I do not understand the story, but thank you for telling it. Your story contained many visual elements, so I was seeing you at numerous moments."

"As long as you're not sending x-rays at me."

"I am not. Did you know Scrooge, or did he live long before your time?"

"Oh, he never lived. It's fiction."

"Fiction. I do not know that word."

"It's not a true story."

"Not true? Then why did you tell it?"

"Because it holds the record for most adapted story, I think. It's one of the few stories everyone knows, even if they've never heard it directly."

"Why would anyone tell a story that is untrue?"

"For fun, mostly. Entertainment. And they can express many great truths."

"Why not simply tell the truth?"

"Nobody likes to be preached to."

"So humans only listen to truth when told through something that is false?"

Stephen took a few breaths. "Not everyone... But people in general tend to listen better when the moral is wrapped up in a story."

"How do you tell the difference between truth wrapped up in lies and outright falsity?"

Stephen considered that. "Sometimes we can't."

"You are definitely not ready to explore the contacted universe. When you no longer have to wrap new ideas up in an untrue story for people to accept them, then you will be ready. As for you, I believe you are ready to experience a little more of this world. Follow me."

She began moving away. Stephen picked up his backpack and followed her footsteps. He ran several paces before he realized there could be things in his way, then he stepped carefully, following her footsteps. It seemed easier now. He could hear how the sound changed when he approached a tree, or a person, and he walked around them.

After the fifth time, he realized what he was doing. The world was there. His ears saw it for him, and Capca's footsteps had become the light. A change in pitch one way meant a tree was in front of him. A different change in pitch meant a person was in front of him, or to his side, and he didn't bump into anyone now. As he followed Capca, he realized he wasn't in an abyss anymore.

Stephen realized Capca's footsteps were gone. He had nothing to follow, but the world was made of sound. People were everywhere, and now their footsteps and wingflaps told him where they were. He could hear the way the sound bent around things, and the trees and plants became real. Stephen felt part of the world instead of being attacked by it.

Several bats approached, clicking at him. Stephen felt them there, wandered around, testing himself and his confidence. He still stumbled on uneven ground, but for the most part he knew where he was and where everyone else was as well.

He decided to try something. He bent down and removed his shoes. Standing barefoot, he felt the footsteps of everyone around him. He felt flapping wings overhead. Hundreds of creatures flying above him. Hundreds of creatures walking by. Stephen couldn't tell if they were looking at him or not, but for the first time, the world had depth, and he did not feel helpless.

Their sounds bent around a tree just to his right. He walked toward it. He felt it coming closer and reached out, contacting it exactly when he expected to. Stephen dropped his backpack and climbed one of the thin branches. He found a higher branch and hoisted himself up. He was searching for something, and he was sure he would know it when he felt it.

Finally he felt a texture that did not resemble part of a normal tree, an extension of the trunk towering far above him, about as thick as his thigh, extremely sturdy, but yielding, one of the streamers that somehow stretched above the clouds.

Stephen tugged on it. It did not move an inch. Stephen wrapped his arms around it and tried to climb it, just like in PT. It had been a while, but he was sure he could do it.

He climbed the streamer until he lost his sense of the world, and the planet returned to the feeling of darkness being a wall. He didn't enjoy this feeling, but he hung on and waited.

A bat flew by. Its wings made a uniform sound for the moment, and Stephen waited. More bats flew around him, more people in the air swooping down, landing, taking off again.

The world expanded. He began to hear how the air bent around the streamers. There were hundreds of them in earshot, all spaced unevenly, and the Heeke flew between them. When they weren't flying near him, he lost his sense of the distance, and the planet returned to flat darkness. Stephen hung here for some time, waiting for people to fly near him and give him another glimpse. He became aware of the different sizes of streamers and the different people flying around him for as long as their sounds reached his ears, and then all became dark again. Stephen wished he could hear farther, know more of the world and what he was missing, but the most he could do was glimpse their world in dark flashes of clarity.

"Stephen!"

Kylac's voice below him. Stephen looked down out of habit and saw nothing but a wall of black.

"Come down! We need your help!"

Stephen began a controlled slide. He couldn't judge how far it was, so when he collided with the trunk, he gasped in surprise. He sat still for a moment, feeling around with one foot. He couldn't find a branch.

"Uh, guys, I'm not sure how to get down."

"Don't panic," Deka replied.

Stephen felt something large next to him. He felt warm breath on his neck, and then teeth gripping his shirt. The teeth hoisted him off the branch. Stephen let go, and his limbs seemed to fall away as he moved down. His bare feet touched the dirt, and suddenly he had arms and legs again. He stood up.

He heard raptor claws scraping together. "Thank you, Capca."

Stephen turned and looked in her direction. "Were you there the whole time?"

"I would never leave you alone in the dark."

She made that throaty sound again. Now he knew it was laughter.

A canine hand touched his back and turned him in one direction. Stephen walked with them, Capca's footsteps next to him, sometimes making heavy thumping noises, sometimes making no sound at all.

6

"I don't like this place," Stephen said. "It's too quiet. I feel like I'm in a void again."

"We're about to open an offworld portal," said the batthing in accented English that tugged Stephen's ears. Stephen thought he sounded like a spastic little kid, but Kylac had told him Lalks was in fact an adult. "Deka and Kylac are going to let me visit a lighted world."

"It will be a taste of a lighted world," Deka corrected.

A reptilian hand reached into the pouch around the human's belt and removed the multi-tool. It slipped into Stephen's palm.

"Deka?"

"Lalks needs to know what your world is like."

"Okay..." Stephen turned it around until he found the switch.

"You shouldn't be doing this," Ollet said from behind them, also in English. "This could kill you, Lalks."

"How would you know? You've never been to a lighted world."

"I have been alive a long time. I know what a lighted world is, and I never understood why you are so eager to go to one of those places."

"I want to help! At least the Relians understand."

"We do," said Kylac. "Now stand there and be prepared for light as you have never imagined."

"Guys, what's going on?" Stephen asked. "Help out here. I'm kinda blind."

"The Heeke usually prepare themselves for lighted worlds by visiting brighter and brighter planets," Kylac said. "Lalks thinks he's ready without visiting such places."

"And Ollet regards him as the son she never had and doesn't want to let him go," continued the raptor.

A high-pitched croaking sound came from behind them. "I lost my child! I told her not to go. I told her she wasn't ready! Everyone knew to stay away from that part of the hub. She didn't even enter the portal. The light coming from it was enough to kill her."

"It killed her?" Stephen said.

"That's not going to happen to me," Lalks said. "I prepared."

"Stephen," said Deka, pulling his hand to the side, aiming his wrist down and to the right a little. "Show him how a lighted world feels. Turn it on."

Stephen took a breath. "I feel like I'm about to commit a serious crime here."

He clicked the button.

On any part of Earth, this light would have been weak, but here it seemed as bright as a nuclear blast. Lalks became illuminated in yellow. His body was bat-like, covered in short, brown fur, and his hands were in fact part of his wings, the fingers joined by a sheet of skin, just as Stephen imagined.

Lalks shrieked and writhed as if Stephen had just shot him with a gun. He dropped to the ground, screaming and writhing.

"Off!" Deka shouted.

Stephen switched it off, and the world became a wall of darkness again.

A large body moved around them.

"That is a weak beam of light, Lalks," said Ollet. Her voice came from directly over him. "A planet will be worse!"

"That..." croaked the bat on the ground. He gasped.

"It hurts, doesn't it?" Ollet said. "Please listen to me! You're lucky the hub hasn't had portals to lighted worlds on it. If you even went near one, you could be dead! Again, Stephen!"

"Again? Are you sure?"

"Again!"

Stephen switched it on. The brown bat-thing was still on the ground, writhing and screaming in frequencies Stephen could not hear but somehow felt.

Stephen caught a glimpse of a tall leg with white skin. Ollet gave a rumbling scream as she stumbled away. Stephen almost moved the flashlight to track the strange leg, but Kylac and Deka held his hand and swung the beam back to the bat.

Now he got a good look at the Heeke's face. The majority of it was ears and nose, and there wasn't even a place for eyes. Stephen didn't wait for a command. He clicked the light off, and the bat's cries ceased. Gradually, so did Ollet's.

"I'm sorry," Stephen said in her direction.

"It was my fault." He heard her catching her breath. "Now, Lalks, that is what electromagnetic radiation does to your body. Do you understand now?"

"Please... Let me start. Let me build up."

Ollet's voice became firm. "Lalks... I won't lose you to the light, too."

"I want to go. Why did I become an Archeon if I can't?"

"Ollet," Deka said, releasing Stephen's hand. "Lalks is old enough to mate. Don't you think it's time to let him begin?"

Nobody spoke for some time. In the silence, the sound of Stephen's own heart beat in his ears became deafening.

It was so quiet he heard blood flowing through his body, which made him nervous.

"It takes years to build up tolerance," Ollet said "Do you understand that you will die if you leave unprepared?"

"Yes," gasped the Heeke. "That was..." He made crying sounds. "All that time letting the Wen see me. It wasn't enough."

"You can't just jump into this," Kylac said. "Take your time. And Ollet. You've been a good mother to him, but it's time to let him explore."

She walked away from the bat, her footsteps uneven. She was limping.

"I'm—I'm really sorry," Stephen said. "I didn't know you were so close to him. Does... light really hurt you?"

She moved closer to him. Stephen felt her breath on his face.

"The Heeke can build tolerance to it. The Wen cannot. My daughter forgot that. Thank you for reminding Lalks of it. You may have saved his life." She paused, turned away. "He is too young to be an Archeon. Far too young."

She limped away. Stephen slipped the multi-tool back into the pouch and buttoned it. He felt as though he had a death ray in his utility belt.

Kylac ran to Lalks, who was still reeling from the shock, and held him. Stephen guessed the fox was holding the bat. "You'll be fine. You'll get better. Come on, stand up."

Lalks was panting and making the same clicking noise over and over again. The noises rose in height. Kylac had helped him to his feet.

"Is he crying?" Stephen whispered.

"Yes," Deka said. "That was the first time he ever felt electromagnetic energy stronger than a Wen's."

"That hurt me just to watch. But they have no eyes. How does light hurt them?"

"It affects their whole body," Deka said. "Just as it does yours."

Stephen could hear Kylac and Lalks. The fox was helping him walk.

A head descended next to Stephen's, and Capca's voice filled his ears. "Just think what all those electromagnetic waves on your planet are doing to you."

Stephen gulped.

"I'm making a way off this planet," Deka said. "We're going to Neben. We didn't intend to take you there, but their Archeon is looking for us."

"Fine by me. I'll miss this place. I was just starting to get the hang of living here."

Deka lightly clawed him down his back. "You were a good ambassador."

Neben

Stephen stood on the sand, reminded of California. The vegetation around him looked stable and healthy, but with the precarious sense of the desert underneath at all times, just waiting to take over when the water dried up. The sun had almost slipped completely below the horizon, and the temperature was beginning to drop.

"There will be no light shortly," Kylac said. "Neben is dangerous for most offworlders to visit during the day. The atmosphere doesn't keep out all the radiation, so this planet is bathed in it."

"You're sure it's safe?"

"At night, yes."

Deka scented the air. "She's here."

Kylac and Stephen walked a few paces behind Deka, feet sinking up to their calves in sand.

"Should I take my clothes off now?"

"Yes," Kylac said, waving his tail. "I don't know why you bother putting them on between worlds. It will get cold here quick, but stay close to me. I have plenty of heat to give."

Stephen chuckled as he removed his backpack and pulled his shirt off while he walked.

"The water will stay warm until dawn, just in case I need to leave. Deka will probably be there all night. During the day, we'll take shelter in the tower, or we'll take a way to the other side of the planet until we're ready to leave."

"Is this place really that dangerous?"

"For us. Not for them."

Stephen saw a faint, green glow ahead. Some of the glow took to the air, flew over their heads, and landed in the hub, ducking through one of the portals that opened into daylight.

"Birds?" Stephen said.

"Their feathers reflect radiation in the daytime. It retains a faint glow for most of the night. The mammals, however, are covered in thick plates of skin."

"Yay, people with skin."

Kylac's tail wagged.

Stephen reflected on the story the Relians had told him about the last time they had visited, how the Nebens found a cave system full of tiny crystals lining the walls and ceiling and even the floor. As they explored, the crystals sent out sparks of electricity and began growing and reaching out to them. The crystals had followed them all the way to the final chamber. Deka, Kylac, and Neben's Archeon just barely made it out before the cave became a tangled forest of cave crystals shooting lightning everywhere.

"Do you think they found another cave? With more of those growing crystals?"

"I hope so."

By now Stephen had removed all his clothes and stuffed them into the bag. He didn't even wear shoes, though he feared stepping on something sharp, as the vegetation reminded him of places where cacti hid.

They entered the midst of the settlement. Plants grew everywhere surrounding a large, finger-shaped lake fed by a portal suspended in midair. Armadillos the size of bears but covered in plated skin walked on all fours around them.

Heron-like birds milled around the water, their green plumage becoming brighter as the twilight dimmed.

"They're beautiful. But—you said they're glowing in radiation? Is it safe to be around them?"

"The radiation is just photons at this stage."

"Cool."

Now the people began to take notice of Stephen, and soon a flock of birds had surrounded them. Some landed in their path, stalking up to Stephen's face and studying him. Noises came from within their closed mouths and exited from a bony tube sticking out the tops of their skulls.

"Is that their language?"

Kylac was making similar roaring and whooshing noises with his open mouth. It didn't sound quite right, but the people seemed to understand.

More of these sounds came from the things on their skulls. The birds did not open their mouths when they spoke. Their faces reminded Stephen of those grotesque bird masks from the dark ages.

Now the mammals joined the birds. They stood about as tall as Stephen's chest while on four legs, and their skin was thick and plated.

"These guys are not so beautiful."

"Back to judging based on sight?" Kylac said, waving his tail.

"They look like armadillos on steroids."

"Pangolins are actually closer to their appearance and habitat."

"What's a pangolin?"

"Never mind. Their skin protects them from the radiation. They can stand all day in it and never feel over-heated."

The crowd had become so thick he and Kylac couldn't move. Kylac made rushing-air sounds at them, and the people replied with a hurricane in return. Stephen wished he could understand.

"They're remarking at how exposed your body is," said the fox. "They want to know if your planet has any solar radiation at all."

"Yes, but that's why we wear clothes. Keep warm, keep the sun off."

Kylac repeated his words for the group as more people poured in through the portals. Stephen wondered if he should have prepared a song, or a play, or a tap dance, or some sort of comedy routine just to give all these people a reason to be here.

2

Deka followed his nose to Qan as more and more people flocked to see Stephen. They were elated to meet a new species, and were confused that his kind was still uncontacted. Deka yearned for the days before the disaster, when word spread quickly between the planets, so they wouldn't have to keep explaining themselves over and over.

A dozen Jilit stood mixed in with the Nebens, just as interested in Stephen as the natives. These green-furred mammals walked on all fours, standing up to Deka's knees at the shoulder. Their snouts were long and pointed, meant for insertion into the ground, where they used their sub-audible voices to detect prey. Over the generations, their sonar had become sensitive enough to find burrows hundreds of paces beneath the surface.

A curious thing about their biology was their lack of ears. Their auditory nerves were in the feet, and they communicated by touching paws to one another's vocal cords. While they could hear voices indirectly through the ground, they preferred hearing by personal contact.

Deka was relieved Qan could open offworld portals again. Judging by the number of local ways at the hub, she must have reestablished contact with the entire planet by now and restored the oases.

Deka lost Qan's scent among the Nebens and Jilit. He raised his snout and took in the air again. He no longer smelled her, which meant she had either moved upwind of him or disappeared into a portal. Deka huffed, turned around, and watched Stephen and Kylac.

Stephen was speaking, and the canine translated. Deka tried to imagine how Stephen appeared to the Nebens, and at the same time he considered the human would be able to identify with them better than anyone he had met so far, as the Nebens were sight-based creatures.

While Deka pondered these things, his nose caught a familiar scent. His entire body shifted to face it, and he centered on a particular Neben bird, glowing bright green against the darkening sky.

She smelled like Sonjaa.

Deka's knees weakened. His breathing stuttered. For a moment his eyes saw Sonjaa there instead of the feathered Neben, and it was only with a great deal of concentration Deka remembered she was not a Relian reptile.

Deka hesitated to approach. He stood still for a few breaths. Stephen's voice filled the air, followed by Stephen and Kylac discussing, translating, and then Kylac's voice addressed the Nebens and Jilit.

Finally Deka stepped toward the bird. Her scent strengthened, and Deka almost cried again. It felt so good to smell her again, and his mind tried to override the emotions.

"Hello," Deka said in her language.

The bird turned and faced him. "Hello, Deka." She had answered in Relian without opening her mouth. Her words had the same echo and rushing air sound as her native language.

"You speak it very well," Deka replied, also in Relian. "You wouldn't happen to be an aspiring Archeon, would you?"

"No, but I do have a talent for languages. I find languages spoken with the tongue instead of the throat more interesting. Mimicking the sounds is a challenge. I enjoy it."

Deka realized he had been moving closer and closer to her. He backed away. "You smell like my mate."

She tilted her head. "I do? What does that mean?"

"It means..." Deka swallowed, started again. "It means... I don't know what it means. You even sound a little like her when you speak Relian."

"Thank you. Is your mate well?"

"I have not seen her since the disaster."

"Oh, of course. I almost forgot. Rel..."

"I haven't given up finding her. I'm sure she's alive."

Stephen laughed, and Deka held his claws together; he had been listening to the joke the human wanted to tell. Kylac was telling him the humor would not translate.

"You should meet Stephen," Deka said. "His language is all tongue."

"I know. I've been listening."

She stepped through the flock of glowing birds and plated pangolins a little more so she could hear better. Deka followed in her wake. He now realized how little she resembled a raptor. She didn't walk like one either, and the way she spoke without opening her mouth looked eerie, and yet he felt as though he belonged with this person. He allowed himself to feel it.

"What's your name?"

She did not face him. "I am Uum."

"Have you been offworld?"

"Many times. I have learned more than three hundred languages."

"And have you ever wanted to know if you can become an Archeon?"

"I'm told I could be with how I'm learning languages like this, but I don't want that life. All those people relying on you, always being called to open new ways to other worlds, thinking about those portals day and night all your life to hold them open. How do you do it?"

"It can be stressful," Deka admitted, lowering his head as he walked, "but I have a talent for it, and I'm glad to use it."

Uum pushed her way through until she was close to the front, Deka following close behind. Kylac was talking just a few words behind Stephen. Stephen was visibly distracted by the running translation, as he kept hesitating and looking at the fox.

They found a place close to Stephen and stood together among the Nebens. Deka watched Uum as she watched the human.

Someone asked him why his genitals were so exposed.

"Because God has a sense of humor," Stephen said.

Kylac translated it as: "Because evolution is a joke." It was the closest the two languages could agree and still preserve the humor.

Someone else asked him how often he needed water.

"Quite often," Stephen said. "Several times a day."

Uum watched Stephen as an Archeon would, dissecting his words instead of merely listening to them. The raptor breathed her air, and she smelled more like Sonjaa every moment.

Stephen took several more questions, answering them in English, but the people were listening to Kylac. Then Uum spoke up in heavily accented English.

"How long did it take Kylac to convince you to have sex?"

It shook Stephen to hear English coming from the audience. Many in the audience turned to look at her, stunned.

Stephen laughed. The Nebens did not know what the sound meant, and they were bewildered, thinking this was more of his language and waiting for Kylac to translate.

"Less than an hour, I think. You speak English?"

"I'm still learning," Uum said.

"Your accent is weird," Stephen said. "I don't even know how to describe it."

He went back to taking questions, and answering through Kylac.

Uum turned to Deka and spoke in Relian. "What did he mean by that?"

Deka rubbed his claws. "You have a very unique way of pronouncing English without lips or a tongue. I think he wants to hear more. I do, too."

Deka's flank bumped up against Uum. He hadn't realized he had been slowly moving closer to her. He wrapped his neck around hers and growled. Uum did not seem to react. Deka growled lower. He ignored the glowing feathers and let the scent take over. Then Deka remembered she was not a Relian reptile, and she might not understand what this meant. He stepped away, leaving half a pace between them.

"No, I can't walk on all fours," Stephen said. "Well, I can, but it's not comfortable, and I can't do much."

Kylac repeated Stephen's words in the Neben language.

3

The cavern glowed in faint blue light and was so large they could not see one end from the other. Small crystals arranged in straight rows covered the ceiling and walls. Dozens of stone buildings sheltered crystals within them. The tower in the center of the cave stood as tall as fifty raptors. The bottom stone had been removed, all the water drained. The entire cave floor was damp.

Neben birds darted through the air but never approached the crystals that lined the walls and ceiling. The mammals admired the view from below. Some milled about at the top of the tower, looking down and around. A few Jilits had also come. Some of them knew the Neben language and held paws up to the necks of mammals, sharing words. One held a paw partially over the hole in a bird's skull, trying to feel the moving air.

Kylac stood next to the tower, gazing up at the crystals on the ceiling. The spinning portal overhead that supplied fresh air to the cavern drew his attention for a moment, and then he followed the line of crystals overhead. Kylac's ears flicked back and forth.

Dozens of insect bodies made of crystal lay strewn about the floor, all appearing to have collapsed in the middle of performing some task. Deka, Stephen, and a pangolin-like creature wandered around the cave, looking at each of them.

"This cave is set up the same way as one we found," Qan said. "I think it's artificial. So far it's the only other cave the Jilit have located."

"Located?" Stephen asked. "How?"

"The Jilit use their voices to find prey underground," Deka said. "I'm surprised they could hear this far down."

"It pushed their limits, but they told us where to dig. We found another tower under the sand, and it connected to another cave system. They also heard how the system was oriented, and I calculated where the main chamber would be so we wouldn't have to disturb any more crystals."

Stephen stooped down and examined one of the crystal insects, six limbs, an obvious head, what looked like antennae poking out the top. No visible joints anywhere, no indication how they would have moved.

"What are they made of? Quartz?"

"We're not sure," Qan said, "but Kylac named it Dekanite."

Stephen smiled wide and looked up at her. "Dekanite?"

"After the Relian reptile who retrieved our first sample."

Deka stood next to Stephen, looked at the crystal insect on the floor, and then turned to Qan. "Were they alive?"

"That's why I wanted you to come. I realized these caves must have been flooded at some point, so I closed the portals over the lakes around here, and the cave began to fill."

"What happened?"

"I'll show you, but first I want you to see this."

Qan walked away from the Dekanite body and toward the other buildings. Deka and Stephen followed her. She led them across the cave floor to one of the shorter structures, one that resembled what they had found in the other cave: a table running lengthwise across the entire building, tools laid out, glass vessels containing tiny pieces of Dekanite, glass cubes on the table, and Dekanite bodies collapsed at each station.

"I think I understand what they were doing here," Qan said. "It's very difficult to make out, but the patterns in the glass cubes resemble the patterns they were carving into the crystals. They were building the crystal, one layer at a time." She picked up a tool that looked like a syringe. "That's what this was for. We found several of them full of solid mineral. I think it had once been liquefied, though I'm not sure how. Once they spread a layer of it on the crystal, they used this."

She held up the one that looked like a scraper and a syringe.

"From what we can tell, they did it one molecule at a time, building the crystal structure layer by layer, and then planting the seeds in the cave to let them grow. I thought, why would anyone do that? Why would anyone create crystal structures? Why not let them form on their own? Then I thought, why run electricity through them? Then I realized—"

Kylac had just arrived and was standing at the door. "They're creating neural pathways."

"Yes, it's how they reproduce. I'm sure of it."

"Wait," Stephen said. "They're etching crystal like a brain? What good would that do? It's crystal. It can't move or think or do anything."

"It can grow," Kylac said. "It can definitely grow. I remember that giant crystal that followed us. When it glowed, I saw patterns inside it similar to the ones in the cubes." He turned to the bodies on the floor. "These Dekanites must have been able to move."

"The electrons came from the water itself," Qan said. "As it filters through the soil and the rock, it picks up stray electrons. Enough to power a life form in here for years."

Kylac leaned on the table, ears turned back.

Stephen was bent over, examining one of the Dekanite bodies at a station.

Deka leaned on the long table, gazing into one of the cubes. The etching was so fine he could only make it out by how it distorted the background.

Qan stood beside Deka and observed the cube. "The etchings are crystalline structures at a molecular level. Neural pathways preserved in glass."

"Whose neural pathways?" Deka asked.

"I don't know. We found where they stored the cubes. It's another building off to the side. There are as many cubes as there are crystals planted in the ceiling and floor."

Kylac's ears bloomed.

A familiar voice came from the doorway. "And their language is incredible."

Deka turned and stared at Uum. She had spoken in English. Her scent filled up the structure, and it became the only thing Deka could smell. Deka glanced at his fox, who was taking in her scent from a distance as well.

"What language?" Stephen asked.

"Qan will show you soon. I tried to learn it directly, but it's too painful. I've tried to observe it, but my eyes are not good enough to see what is going on. Being down here in the caves when they are active is dangerous."

"What do they do?" Kylac asked, leaning forward, vigorously scenting her now.

"I'll let you see for yourselves," Qan said. "I have stopped the spin on the portals to the lakes in this area. The cave will begin to flood shortly. I will only let it go up to our waists."

Kylac had been slowly walking forward, past Deka, meeting up with the Neben bird, breathing in more and more of Uum's scent. Finally he paused to breathe and straightened up.

"Who are you?"

"My name is Uum. I enjoy languages, and Qan asked me to come down here and try to interpret how the Dekanites speak to each other. I am frustrated. I hope the two—" She glanced at Stephen. "The three of you can help."

"One more thing before the cave floods. Come and see this."

Qan walked out of the laboratory. Kylac followed her, pausing to sniff Uum again, meeting Deka's eyes for less than a breath as he walked out. Stephen followed, leaving Uum and Deka alone. They stood still, staring at each other.

Uum spoke in Relian. "Deka, I visited Rel many times. I recognize your body language."

Deka took a step back, bumping his leg against the table. He suddenly could not meet her eyes.

"You smell so much like Sonjaa. You even sound like her, especially when you speak Relian. I'm not sure what it means, but... When I'm around you, I... I found her again. It's over. The worry, the waiting, the disaster. I'm sorry. I miss her so much."

Her plumage relaxed. "Are you sure she's still alive?"

"Kylac is trying to convince me the odds are too small, and I may as well find another mate. As if I can, now that... Now that my species is dead."

"You don't know that." Uum moved closer to him and wrapped a wing around Deka's neck. His heart sped up. Somehow that reminded him of Sonjaa as well.

"I know you're hurting, but I'm not your mate."

Deka lowered his head. "I know. I know it in my head, but everything you do... It reminds me of her."

She brought her other wing up and wrapped it around Deka's neck as well. The raptor recognized it as their gesture of friendship. "I wish I had known her. We have work to do. When the cave floods, you will understand what we're dealing with."

Uum stepped away, letting her wings slide away from Deka. He turned and followed her outside. Uum spread her wings, flapped a few times, and took off. Deka tucked his arms into his chest, about to take off running, but he held back. He snorted her scent from his nose, trying to forget.

Qan and the group were nowhere in sight, but Uum landed at the base of the tower. She walked inside as she folded her wings. Now Deka kicked off and ran full speed across the damp floor, tiny drops of water pelting his hide. Deka didn't care about the tower, or the crystal creatures on the floor that bore his name. He wanted to be with Uum again.

Deka entered the base of the tower. It was damp in here, and everything was perfectly intact. The scents of the others led upstairs, and Deka carefully climbed to the next floor. Dekanite bodies lay everywhere, all quiet and lifeless.

Qan, Kylac, Stephen, and Uum stood on the fifth level. Someone had cleared a path through the Dekanite bodies, and the group was examining the walls, glass cubes stacked against them from floor to ceiling. Stephen was staring into one of the cubes.

"I can't see anything," the human said. "They just look like solid blocks of glass to me."

"For a visual species," Uum said, "your vision is pretty lacking."

"I know. So what am I looking at?"

"The glass is etched internally with a fine pattern. At first we thought it must be a written language, but after Qan allowed the cave to flood, we had to adjust everything we thought we knew."

Deka joined the group and stood as far away from Uum as he could, though he made sure nothing blocked his view.

Kylac turned, scenting around. "So they're a noncompatible species with an untranslatable language. Their perception of the world will be so different from ours."

"And all we can do is shrug?" Stephen said.

"Sometimes it's painful like this," said Qan. "You go through so much effort to discover a new species, but you can't understand who they are."

Kylac left the group and met Deka. He locked eyes with the raptor, silently telling him to follow, and then he climbed the stairs. The two ascended, winding their way up the wall, past each level, until the reached the roof. Up here, the blue crystals shined down on them, and the uniform pattern they formed was hypnotic and beautiful. A light rain fell from these cave crystals.

"Kylac?"

The fox stood with his muzzle turned up to the ceiling. He stared at the crystals, ears folded back.

"If I closed my eyes, I would forget she's not Sonjaa. I have no idea what it means, but I have a bad feeling."

"I want to scream! I want to tell her how long it's been and how much I yearned to be with her again!"

"She's not Sonjaa," Kylac continued. "I have to keep telling myself that, too."

"She's so much like her, not just her voice but everything!"

"Get to know her, but don't scare her. Something strange is happening." He paused, following a row of crystals with his eyes. His ears flicked. "Both towers were once full of water. Most of the bodies were found in the towers. Crystals were planted in the walls and ceiling. These people lived underwater. The electrons in it would have sustained them for years. But doing what? Those crystals can grow, so it's likely they will grow up into the Dekanites we found on the cave floor, but these caves aren't big enough to hold that many fully grown Dekanites."

"Reproduction is the only explanation."

"With the information we have," Kylac said. "We need more."

The trickle of water droplets from the ceiling increased to a light shower.

4

Precipitation fell from the ceiling. Deka and Kylac stood outside, observing where the water emerged, how it covered the tiny crystals, and how each drop made them glow slightly brighter. They tasted a few drops as they fell, and sure enough stray elections were in the water, and it was warm.

It quickly became a hard downpour, and the Relians retreated to the shelter inside the building with the workstations. Only a few Nebens remained in here, including Uum. Rain in a cave was such a strange thing to witness, and the downpour increased until the water poured from the ceiling in sheets.

The water level rose to their ankles. Quickly it came up to their thighs.

"The portals are spinning again," Qan said.

It reached their waists, and the flow slowed to a trickle. The birds and mammals remained in the laboratory, as close to the walls as possible, the water only up to their stomachs.

Kylac began swimming to the door, but Qan turned and blocked him.

"Stay in here. It won't be safe out there when they wake up."

Everyone remained quiet. The Nebens around them gazed down at the Dekanites at the work stations. The water had submerged the tools, the glass cubes, and the little jars of Dekanite as well.

Deka noticed Uum staring at one of the Dekanites, waiting eagerly. The theropod sloshed through the water and stood next to her. The water rose, and everyone's eyes were on the crystal people. Deka watched Uum, breathing her scent, letting it remind him of life before the disaster. The water reached Deka's thighs, and the Dekanite bodies on the floor began to glow bright blue.

The water itself seemed to hum with electricity. At another workstation a little ways down the stone table, Stephen was gazing at a glowing Dekanite, reeking of that wonderful scent of fear and wonder his species gave off when encountering something unfamiliar.

Moments later, the Dekanite body near Uum shifted. The limbs moved. The body glowed with electrons, vaguely hinting at the crystal structure inside it. Deka's killing claws rose out of instinct.

A spark of energy pierced the water and jolted through the bodies around the table. It just barely missed Deka, and he sloshed backwards through the water to be near the wall, as the other Nebens had already done.

A blue spark shot around the table. Then another one, white this time. Then a continuous beam streamed around the Dekanites, connecting them. A bolt of electricity came through the door, joining the Dekanites inside with the ones outside, and they began to climb to their feet.

Kylac peeked through the opening. Outside the building, in the cavern itself, the Dekanite bodies also glowed blue, and they were rising up to all six limbs. Streaks of lightning shot between them, reaching all sides of the cave, touching the edges of buildings, feeling for other Dekanites. When they found one, the energy stayed on it, and it became part of the electric web.

Everything under the surface became webbed in electricity. The bolts of energy changed colors and thickness faster than even a Neben's eyes could see, and they touched every crevice of the cave.

Finally all the Dekanites had awoken. The ones at the workstations stood at the tables, staring off into space. Deka peered at the closest one. It had no eyes, no hands, no apparent joints. The crystal itself appeared to be flexing and moving.

Suddenly the electricity exchange between the Dekanites beneath the chest-high water stopped, and the Dekanites began moving in unison. Each Dekanite at each workstation rose up to its two hindmost limbs and reached toward both tools at once. Minuscule sparks of energy emerged from the ends of their forelimbs, wrapping around the tools and lifting them.

Outside, the Dekanites moved in straight lines, single file, some up the cave, some down. Others stood still and stared off into space. A few Dekanites stood at attention by the door, watching, their blue glow dimmer than the more active ones.

Deka carefully approached the working Dekanite. Its head pointed at the glass cube on the table, projecting thin streaks of electricity at it, and it was working the tiny fragment of Dekanite in the jar before it. It held the tools in phantom limbs made of electricity, their movements barely perceptible.

Deka wished he could smell these creatures, but he had a feeling he would not smell much but electricity. He leaned closer.

A bolt of electricity emerged from the highest point on the Dekanite and struck the raptor on the snout.

"Skree—!" Deka hopped backwards. The Nebens around him laughed.

The Dekanite's head turned and faced Deka directly. The raptor froze, meeting what should be its eyes, but it had no eyes at all. No mouth, no face, just solid crystal.

Bolts of electricity shot out from the creature's body and struck Deka on his fingertips, his nose, and his toe claws. The shock sent Deka to the wall, among the Nebens, who were still laughing at him. Deka also heard the distinct sound of a canine's tail smacking the water. Just as Deka was calm and still, a human shouted and splashed.

"Shit! Shit! Shit, that was my dick! Shit, that hurt! Shit! Fuck!"

The Nebens laughed harder as the human stumbled backwards, holding himself.

Everyone stood against the wall now, afraid to go any-where near the Dekanites, except for Uum. She leaned over the Dekanite at the station, dipping her beak below the water, as if trying to force it to shock her.

Green energy streaked from the Dekanite body and struck Uum. She groaned, bent at the knees to hold her stance, and did it again. Another bolt struck her, ruffling her feathers.

Deka wanted to pull her away, but he held still as Uum reached down and tried to touch the crystal life form, but as soon as her wing came too close, energy reached up and coursed through her wing. She absorbed a lot of voltage before backing away, shuddering and catching her breath.

5

"Please tell me what I'm looking at," said the human.

Kylac was standing closest to him, and he answered. "Can't help you this time. This is new to me, too."

The small group stood at the top of the tower. It was the best place to observe the Dekanites around the cave that were moving to an invisible beat along the straight paths worn into the stone.

Crystal people, all glowing bright blue, moved here and there. Others stood in formation in front of the work-station buildings, or in front of the cave mouth, or off to the side on the far end of the cavern, arranged in straight rows.

One individual in this the group of people standing idly in front of the building received a bolt of energy from inside, broke formation, and entered the structure. It emerged a moment later with a glass jar in its electrical fingers. It carried the jar in the direction of the caves, merging with the matrix of Dekanites already there, and waited, still holding the glass jar. Its glow slowly faded.

Another Dekanite walked into that structure and emerged holding a glass cube. It followed a straight path to one building at the edge of the cave, passed the cube to another Dekanite waiting there, then walked to another structure and emerged holding another cube. It walked along the paths to the workstation, entered, and came out with nothing in its hands. It then stood in formation with the others and waited.

There were almost a hundred Dekanites in the cave, but most of them stood in rigid formation off the paths, their internal color dimmed, while the others performed actions around them.

"They're like ants," Stephen said. "How are they alive? How are they moving?"

"Anything is possible," said Kylac, "but this is incredible. What kind of environment would produce living minerals?"

"Possibly the same one that gave rise to the life forms of Towe," said Qan.

"But those people still have organic parts," Kylac said. "These have none. Have you let the Selts examine any—oh, no, of course not."

"The Selts are just as helpless as we are," Qan said.

"Their language is right in front of us," said Uum. "They live off the electrons, and they use the electrons to speak to each other. I've been trying to understand it, but it's too painful. Can any of you make out any patterns in their speech?"

"I can perceive the flowing electrons, but the patterns make no sense to me," said the raptor.

Kylac spoke up. "Even the sound it makes... There doesn't seem to be any pattern to it at all."

Stephen turned to Kylac. "I thought you guys could find patterns in anything."

Deka smiled with his hands. "This is a little beyond our senses. We're aware of the electromagnetic energy, but not always the content."

"Otherwise we wouldn't have needed a television to watch PBS," Kylac finished, bumping Stephen's hip.

"That is as far as we have come with the language as well," Qan said. "Those Dekanites are idle because their stations are inaccessible right now. Once the water level is high enough, the others will activate and swim into the caves, or to the ceiling."

"Kylac," Stephen said. "I thought you and Deka said it wasn't normal for people to use tools or build stuff."

"For lone species, generally," Kylac answered, "but it's too soon to judge."

Multicolored streaks of lighting occasionally passed around the Dekanites in the cave, even the idle ones. They did not seem to notice.

"The voltage must be immense for the electricity to do that through water," Stephen said. "A nine volt battery to the tongue is bad enough."

Deka rubbed his claws. "How's your dick?"

"It still hurts!" He laughed.

Water began to trickle down from the ceiling. All eyes turned to Qan.

"I will let the water rise higher so you can observe more of them in action."

The drip changed to a shower, then a torrent, but nobody on the tower retreated for shelter. Everyone watched the Dekanites as the water rose. More and more bolts streaked through the water, connecting every crystal creature in a lightning web.

The rain slowed to a trickle when the water had reached halfway up the tower. With one more web of lighting, a few of the Dekanites who had been standing in idle formation suddenly glowed bright and began paddling above the level of the others. They paddled up to the height of exactly one and a half Dekanites high, and swam for the caves, slowly kicking the broad paddles at the ends of their feet.

"I don't remember those," Deka said.

"That's because none of them had paddles on their limbs until now," said Qan.

"They can change their bodies?" Stephen said.

They paddled in a line to the cave entrance, now just barely underwater. The inside of the cave was visible from here. Deka and Kylac peered closer as the swimming Dekanites approached it.

The crystals planted around the cave glowed, but the ones underwater glowed brighter. Five Dekanites swam through the entrance and floated under the surface. Lightning bolts streaked from the crystals on the wall, floor, and ceiling and poured into the Dekanites. Each crystal fired of stream of electrons concentrated on one of the Dekanites in the center of the chamber.

"They're not growing," said the fox, his fur soaked and clinging to his lanky frame.

"Good," said the raptor.

"They were reaching out to us," Kylac said.

"Neural pathways etched into crystals," Qan said. "Each of these seeds was a conscious mind. A mind without a body, existing in total darkness, unable to see, taste, touch, or hear anything until we arrived. When they felt us, they sensed something for the first time, they became aware of an outside world, and they wanted to reach out and touch it."

"Imagine living trapped inside your own thoughts," Kylac said to the human. "Then with one spark, you realize you're not alone."

"Those Dekanites could be caretakers," said Qan, watching them channel all of the energy from the crystals. "Teaching the young, helping them learn, giving them stimuli so they don't lose their minds in sensory void."

"We did disturb them." Deka gulped. "I may have killed a few when I broke them off the floor."

The energy flashed, and each crystal was saying something different—the energy they gave off varied in color and intensity.

"But there are so many of them," Stephen said. "Where would they live?"

"I have no answer for that yet," said Qan.

The caretakers continued talking to the young. The five of them were joined to every crystal in the distant cave in a web of electricity that did not seem to cycle through any colors at all.

After a few minutes of exchanging bolts with the crystals in the nearest cave, they attempted to swim up to the next cave, found it dry, and paddled back into the main chamber. They returned to the formation, descended into their positions, dimmed, and held still.

"There's no water in those areas," Qan said. "I'd have to flood the whole cave to fill it all the way up. I'm sure the rest would become active then, and the caretakers would be able to speak to every seedling in the system."

The Dekanites continued to move about the cavern. The ones in formation in front of the workstation building walked in and took jars of crystal. Others removed glass cubes and stored them in the buildings around the edges. Others came in with new cubes. Still others came from the bottom levels of the tower and provided new crystal seedlings. The largest group of idle Dekanites, the "caretakers," floated idly, waiting for the water level to rise.

Eventually ten Dekanites waited at the cavern opening, holding pieces of crystal in jars. Everyone in formation at that position had a jar, and the group sent a streak of lighting to one Dekanite nearby, who passed it to others, then to others, and all around the cave.

The Dekanites slowed to a halt and dimmed. The cave became quiet and still.

The people on the tower watched and waited, but nothing more happened.

6

Qan had created a new portal at chest height at the center of the cavern, gave it a spin, and removed some of the water. Nebens wandered around the cave, observing the Dekanites. If they had not observed them moving about earlier, they would never think these crystal figures capable of motion.

Stephen, Kylac, and Qan went back to long building with the workstations, observing the idle Dekanites. They had not collapsed on the floor, but rather held position at their stations, waiting for something.

Out in the cavern, Deka and Uum stood out in the chest-deep water before the large group of Dekanites who had been idle this whole time. The crystal people were the exact same size and color, and held still in exactly the same position.

Uum reached out to one. A weak bolt of energy flew from the Dekanite and shot through her wing and down to the floor, and the crystal creature turned its head and observed her. It sent more bolts of energy her way, ruffling Uum's feathers. She did not back away, but held still and tried to take it. She absorbed the voltage the same way she had absorbed Stephen's words. Eventually she backed away, cringing and exhausted. The Dekanite turned its head and resumed its statuesque pose.

Deka sloshed over to her, lay a hand at the base of her neck. "Anything?"

"It hurts too much."

"They might not even be aware of beings that are not like them. Electricity is the only way they sense the environment, and if they are a lone species, they don't know how to understand anything that is not like themselves. We may be invisible to them."

"There has to be a way to reach them! I'll figure it out. I just need more time."

"I wish I could agree with that."

Deka's flank touched Uum's. Again, he hadn't realized he had been moving closer and closer to her, which perturbed him; as an Archeon, he was normally aware of everything he did. She stood upright and faced Deka. She rubbed her body against Deka's, and the raptor did not smell surprise coming off her.

"Deka... I have a strange feeling."

"What is that?"

"Does the Ya'mah portal need adjusting?"

Deka glared at her.

She shook her head, confused. "What does that mean?"

"That was the last thing I said to Sonjaa before the disaster."

She held Deka's gaze. Deka did not turn away. He fought an overwhelming urge to embrace her and never let her go. Her scent resembled Sonjaa's even more now.

"I'm sorry," she said.

"Is there anything else, Uum? Please tell me."

"No. It's just a strange feeling. I hoped you could tell me where it came from."

She stepped up to him, rubbed her neck against his. Her feathers felt like smooth scales.

"That feels familiar," she said. "I've... I've never been this close to a Relian. I've never done this before. Why does it feel familiar? Why do I have the feeling I know you?"

Deka rubbed her neck with his in return. "Trust your subconscious."

She backed away, regarded Deka, and then spun around. She slowly walked up to the formation of statues. Deka followed a few steps behind her.

"Sonjaa, I—" he caught himself, lowered his head.

"I have lived on this planet my whole life. I am not your mate. I do not want a mate."

"That's... That's what Sonjaa said to me a few days after we met. Uum, you are Sonjaa. I don't understand it, but you feel it, too."

"I don't know what I feel."

"Uum, there's something I need to tell you."

She halted at the lineup, hanging her head.

"Sonjaa was good with languages, too."

"She was?"

"Yes. Very good. I kept telling her she should be an Archeon, but she didn't want to be overwhelmed with so much information at once. She couldn't imagine how she would think, or do anything, so she focused on languages. Not just speech, but how language shapes the understanding of concepts and influences how a species comprehends reality, as well as how a species' perception of reality shapes their language—this bidirectional dynamic fascinated her. She always said you can tell a lot about a people by how they speak."

"Deka, I have been interested in languages since I was young. I enjoy it. Before the portals collapsed, I used to travel everywhere, learning the languages of everyone I met."

"I know, I know, but I had to tell you. I think you need to know. It's not just scent and voice."

"I just met you."

"I think you sense something is wrong, too. Please, Uum. Let's go back to the surface and talk about this. Tell me more about yourself. Tell me what you remember and about all the places you've been. I'll speak to you in any language you like."

After a lengthy silence, she turned her head and met Deka's eyes.

"You always enjoyed that, didn't you? Speaking to her in whatever language she was in the mood for. Some languages were good for expressing distress. Others were better for calm moods."

Deka's heart raced and he couldn't catch his breath.

"And she liked having someone to practice with," she continued.

Deka closed the distance. "This is more important than the Dekanite language. Let's return to the portal."

Her eyes were empty. As she turned around to face Deka, her tail feathers brushed one of the Dekanite statues. Current jolted through her, jumped to Deka, and exited out his tail and into the rock. Deka screeched. Uum whooshed.

The statue turned its head and took a step toward her. It sent another bolt at her, which jumped to Deka. Both of them leaped back a few steps. The Dekanite walked forward three steps and more energy their way. It connected with Uum's beak, and she made more rushing noises of pain.

Deka grabbed her and pulled her away. The Dekanite continued walking forward on all six legs. The raptor tried to move faster, but he was slow in the water. The crystal body neared, shot another bolt at them. Both Deka and

Uum cringed, and he pulled her a few more quick paces through the water.

"Deka, it's following us."

The raptor's mind raced. Connections formed, and he had an idea. "Maybe it will follow us to the portal. We can show it the surface. It might react to that."

"To the surface?"

"Maybe if they see it, they'll notice us."

"It could work. How do we convince it to follow?"

"Keep walking. It might have to shock us a few more times, but it will follow. Qan's portal is on the other side of the tower—it's not far."

Deka and Uum walked backwards through the water, with the Dekanite crawling toward them. Deka had calculated exactly how close they had to be when it shocked them, and he kept Uum walking at just that distance.

This drew the attention of the Nebens. The mammals swam up to watch but held their distance. Birds flew overhead, some landing off to the side and swimming in the water.

The Dekanite seemed to lose interest, so Deka held both of them still just long enough to be in range, and then another bolt shot through them. The crystal person adjusted course and crawled after them with renewed interest.

This happened four more times on the way to the portal, and each time Deka and Uum shared a bolt of lightning, the creature resumed following them as if it had done so this whole time.

"Why is it interested in us?" Uum said, air rushing and wheezing through her skull.

"I don't know, but it loses interest exactly fifteen and a half breaths between shocks. It never varies."

"What does that mean?"

By now the portal to the surface was in sight, and Deka veered toward it. Kylac, Qan, and Stephen stood just off to the side of the sphere, watching them.

"What are you doing?" shouted the fox.

"Can you communicate with it?" Qan said.

"Not yet," answered the raptor. "We're taking it to the surface. Maybe if it becomes aware of something outside the cave, it will become aware of us."

Nobody said anything, and nobody dared venture too close. Deka timed the last shock to the final step before they would go through the way. The crystal life form shocked them, pursued with interest again, and Deka turned around and leaped through the way. Uum followed. The Dekanite crawled into the sphere and emerged on dying vegetation.

The water in the lake had fallen, and the plants had begun to lose their battle with the desert. A lot of water had to drain into the soil to replenish the caves, so the lakes and gardens in this area were suffering.

As soon as the Dekanite touched the soft ground, it sent out probes of lighting in all directions that dissipated into the air. It turned around, bending its segmented body into all sorts of shapes.

Nebens and a few Jilit near the dry lake gathered to watch, leaning backwards whenever a streak of electricity snapped the air. Before their eyes, its feet changed from cylindrical stumps to something flatter, more suitable for walking on soft soil. Now it didn't sink as much as it turned around and probed in all directions.

Uum cringed as a bolt shot near her. "Deka, what's it doing?"

"It's never been outside a cave before. It's searching for the walls."

Several bolts struck a few Nebens, and they backed away. One bolt hit Deka and jumped to Uum, and they both moved away from the panicking crystal being. Its glow had dimmed considerably since it had left the water. Having found Deka and Uum, it sent out several probes directly to them, turned its whole body to face them, and approached. It sent several streaks out to them, and the two continued backing away.

"This doesn't feel like curiosity anymore!" Uum shouted.

"It's still panicking!"

"Go back to the portal!"

Deka and Uum led the crystal creature around and back toward it. Most of its shocks went to them, but it still sent bolts out from its entire body, striking Nebens as far as fifty paces away. Everyone ran from the lake.

Deka and Uum turned and jumped through the way. Suddenly they were up to their chests in water. The Dekanite crawled through and instantly glowed brighter. Its panic bolts found other Dekanites, and they transmitted those streaks to other Dekanites, and then to others, and others, all around the cavern.

The Dekanites glowed to life and moved from their statue poses. They sent feelers of electricity all about them. When they found a Neben, they launched bolts of energy at them and charged them, gaining speed. Some of the avians took flight, but bolts from Dekanites shot them out of the air, and they fell screaming into the water.

Every Dekanite in the cavern was now active and approaching the Nebens. The entire cave floor became webbed with energy. Nebens in the water stood still, paralyzed. More bolts streamed out of the water and struck birds in the air.

Uum turned in place, looking everywhere, feathers ruffled. "What are they doing?"

Many of the Nebens were trying to appeal to the Dekanites, but they were met with streaks of purple and blue lighting.

"They're not—" Deka screeched as a web of energy struck him on the chest and nose. Uum struggled to hold him upright, also reeling from the shock.

Nebens all over the cave swam as fast as they could, the birds trying to take to the air again. Some birds were trying to carry the mammals, but they were not strong enough, even in groups of four, so they merely helped them swim faster.

"I can reach them," Uum said. "I can reason with them."

She broke away from Deka's arms, flapped her wings, and lifted off.

"Uum, you can't talk to them!"

"I can stop this!"

Nebens swam to the tower and funneled into the portal behind it. Mammals swam between Deka and Uum, and birds landed in front of Deka as well. Deka could just barely see her now, flying up to where the crystal people had gathered.

Qan was nearby, trying to speak to them. They focused a lot of energy on her. Qan stood in the water, paralyzed, energy streaming from the Dekanites into her hands, out her feet, and back to the crystalline people.

A large streak of energy raced around the group of Dekanites, and then one of them threw it at Qan. It struck her with enough force to knock her backwards. The portal wavered but held its shape.

Deka tried to run through the water, but he made no progress because the Nebens were all swimming against him. Birds and mammals pushed past him, sending waves of water over Deka's head.

Several birds had grabbed Qan from above and were helping her swim toward the portal. Streaks of lightning shocked them, and they fell from the air and settled for swimming with Qan instead, pulling one another away from the Dekanites.

Deka looked over his shoulder. Kylac was helping Stephen swim to the portal. Every now and then the web of lighting paralyzed them, and they keeled over. Then they picked each other up again and continued.

Uum now stood in the water before the Dekanites, and they directed enormous amounts of energy at her. She stood still and took it.

"Uum!" Deka pushed forward. "Get out! They can't understand you!"

She approached the group, taking the pain. Finally she was close enough to wrap her wings around the head of one of the Dekanites. Energy ran through her and danced on her feathers. She didn't make any noise at all.

Deka tried to push through, but too many Nebens were swimming around him. Finally the Dekanites stopped firing energy. Their glow became faint—they had probably used up most of the energy in the cave water. Uum stood still, holding the crystal person.

"Deka, help me!" she shouted.

Deka attacked the water, screeching, trying to part the mass of people escaping the chaos.

"Deka, what's happening?!"

"Sonjaa, I'm here!" Deka screamed. "I'm here!"

"Friend! Rive! Help me, please! Someone! Help!"

Uum was gone. In her place appeared Sonjaa's body, clawed hands replacing wings, scales instead of glowing feathers.

A flash of energy streaked through Sonjaa and the water and the Nebens fleeing the cave, and she vanished with it.

The crowd thinned out at last. The Dekanites fired lightning through the water and struck Deka in the chest. It thrust him backwards five paces. Kylac was behind him, caught the raptor before he tipped over, and led him away.

"Kylac! I saw her! I saw her!"

"I saw her, too, but come on!"

Kylac led Deka to the wavering portal. Deka tried to run faster, but this was as fast as he could go in water as deep as his chest.

Deka and Kylac were the last ones to the portal. When they paused and turned back, they saw the Dekanites had faded almost completely. A few of them had collapsed, and the others were watching the Relians. Faint streaks flowed back and forth between them.

Kylac led his raptor through the sphere. After they passed through and collapsed on the sand, the portal wavered for a moment, and then it closed.

7

Deka, Kylac, and Stephen sat on the sand in front of the lake. Qan also sat on the sand. The lake had not refilled, so nobody could be in the water yet. Word had spread around the planet quickly, and all four thousand Nebens had gathered at this oasis.

"It could have been our reluctance to share electrons," one bird said. "They might not understand we can't speak the way they do. They would be offended we didn't reply."

"They must have understood the differences," said one of the mammals. "They knew we were not like them, so they should have considered that."

"But they have not discovered portal physics," said another mammal. "A lone species may not comprehend such differences. They would assume everyone else in the universe is just like themselves."

"If sharing electrons is how they speak, what would give them the idea to use their voices to attack?" said a bird.

"They did not leave the cave," said someone else. "Their intent was not to kill us, only to force us to leave."

"Why?"

"Could they have been afraid we would harm their young?"

"And what about the metal and glass? You can't make those underwater. Can they?"

"They could survive out of the water for short periods. They must have been able to in the past. They have to know about us."

Deka growled. He lowered his head, trying to shut out the voices, but his Archeon mind took them all in whether he wanted to or not.

Stephen sat, huddled into himself, holding his knees close to his chest. "Kylac."

The fox met his eyes.

"What's going on? What are they talking about?"

"I'll fill you in later."

Deka growled louder as someone else in the herd spoke.

"They should have realized they needed to be more patient. A dialogue couldn't form in a day."

"They do not know portal physics," said someone else. "It would not occur to them to try to perceive the world as someone else does. They considered us a threat."

"But how? We can't harm them. They figured out they can harm us."

"We should go down there and try to begin anew."

"We can't understand them. They can't understand us."

"A noncompatible species living on our own planet."

"Isolated too long. Their first reaction was not to learn more about us, but to force us to leave."

"Why would they do that instead of trying to learn about us?"

Deka jumped to his feet, screaming. When he ran out of breath, he scanned the birds and mammals. The Nebens had fallen silent.

"Of course they attacked!" His voice was halfway between crying in despair and roaring in rage. "They figured it out! You took the water from underground and made the gardens all over the planet! You changed this planet from desert to oasis! You're the reason the caves are dry! You're the reason their civilization ended!"

The silence lingered.

"I agree," Qan said. "That might explain what the towers were for. They built them to hold water when the caves went dry, trying to wait until they flooded again. Now they know why the caves failed to flood."

Deka snarled at everyone. "Uum is dead, and those people down there are going dormant again unless you shut those portals off!

"We can't just go back to the way things were before the portals," Qan said. "There must be a way to reach them. If only we could calm them down. Help them see who we are."

"Uncontacted species," Deka said. "No knowledge of portals. They don't know how to comprehend things any other way. To them, you are the society that ended theirs."

"We could help them," someone said, "if only they had let us—"

"They don't want your help!" Deka screamed. "They're angry! Yes, you a threat! They want you out!"

Everyone was silent for some time, and then the debate changed to how they could reach them. How they could present themselves apologetically. Deka turned away and faced his fox.

"Kylac, are we ready yet?"

He answered in English. "A few more hours."

"Good. We'll take Stephen somewhere better than this."

He stalked away from everyone, killing claws up, still growling.

Stephen reeked of fear. Kylac stood up, held Stephen's shoulder for a moment, and then trotted after his raptor. He sank into the sand up to his knees in places, but Deka was walking slowly, so Kylac caught up.

"Deka, I'm so sorry."

Deka dropped to his knees and cried so loud the Nebens stopped debating. Kylac knelt beside him and held him until the daytime star rose and Kylac was ready to open the way.

Proxima

Stephen sat in the dim red light. The atmosphere was so thin it did not distort the view of space. The human could see every star, and he thought he could see every star that had ever existed. The tiny pool of water at his feet reflected the entire universe back to him.

The plants around him were so minimal they barely counted as alive. The star gave off so little energy the plants didn't make enough to grow leaves, so they remained simple stalks poking up from the ground.

Stephen felt glad to be fully clothed for a change. Kylac sat to one side. Deka sat on the other, gazing into the water, his dark blue scales appearing off-black here. The red stripe running the length of his muzzle and down his back seemed to shine and hover above his body in this light.

The fox pointed to a particular spot in the sky. "Look there, Stephen."

"Is that it?"

"That is your planet's star."

Stephen laughed. "I'm not impressed."

"Rel's was no different. This far out, all stars are the same."

"This place is beyond words. It's like there's no barrier between me and the universe. You can come here and be alone with it."

"Every Archeon has a world like this," said the raptor, not looking up from the pool. "And we keep it to ourselves.

When you become an Archeon, you can't help but feel a connection with the entire universe. Places like this help us focus on it."

"What do you mean connection?"

"He means when you learn how to perceive the universe as we do, you begin to feel as if you're part of it, not merely living in it."

"That's an old movie cliché. Always rolled my eyes when I heard it."

"Try to imagine it," Deka said. "We can make connections between any two points in the universe. We spend our days and nights keeping those connections open, keeping them lined up with the movements of the universe. It does feel as though we're one of those forces that keeps things in motion, like gravity and..." He hesitated. "And electromagnetism."

"Feel privileged, Stephen. We have never shown this world to anyone else."

"Anyone?"

"Never. This is where Deka and I come to be alone with the universe."

Stephen scanned the sky for a moment. "Thanks for showing me."

"We didn't intend to, but Deka needed to come here."

"You should have made a way to another world," the theropod said. "Friend is still out there, still destroying planets."

"You needed to recover. This is the best place to come."

They were silent for a while, each taking in the view of the universe. Stephen and Kylac scanned the sky. Deka stared into the water. The absolute silence of this world focused the eyes, made them sharper, until the entire universe seemed to come into view all at once. Stephen drank it in.

"Even when I was in the Army, out in the field, I never saw the sky so clearly. I wish I could feel what you guys feel. I really do. It must be amazing."

"It's not for everybody," muttered Deka. "There are many times I wish I wasn't so aware of where everything in the universe is, and how it affects everything and everyone."

"I don't see a downside to it. You can go anywhere you want, do anything you want, remember everything, learn anything."

"Some days I would like to give it up," said the raptor. "I never felt I gave Sonjaa enough of me. My mind was always busy with ways to maintain, languages to remember, people I needed to check in with, other cultures I wanted to experience. Every planet I kept a way open to, I was aware of its location at all times. Not to mention being so aware of my environment. Mind always calculating temperature, wind speed, planet size, strength of gravity, rotation speed, what elements the tree is made of, its chemical composition, rate of radioactive decay, and on and on. My mind takes in all of it at once, and I can't ignore any of it. Sonjaa, Rupi, and the hatchlings often seemed like five extra variables in all of that. I didn't want to think of them that way."

Stephen sat silently for a while.

"It was only because of me he became an Archeon at all," Kylac said. "I was the one who wanted to explore the universe and know how it felt to be part of it. But I feel that way, too, sometimes. The mind was never meant to live with this kind of perspective. It's adapted to view basic survival as fulfilling. Mating, eating, making a difference among other people. For some species, it's protecting territory, or making the biggest burrow. When your perspective on the universe is large enough, these actions become insignificant variables. It takes away the satisfaction, and it can leave a hole in your life. Some Archeons never get used to it."

Stephen sighed. "So why do you do it? What's in it for you? What's in it for everybody?"

"Freedom," Deka said.

Kylac rubbed Stephen's back. "We often come here to remind ourselves of that. Archeons give freedom to the people all across the contacted universe."

"Freedom from what?"

"Freedom to live," said the raptor. "It's easy to forget just what our role in the contacted universe is. How much of a difference we actually make in people's lives."

"But on all the planets I've been to, it doesn't look like anybody's living. What does everyone do all day? Don't they have to work? Don't they do anything for fun?"

Kylac and Deka laughed in their own unique ways.

"That is what portals are supposed to do," Deka said. "When two species reach the point where they can understand the universe as it is, they use it to lessen the burden of survival for everyone, ending it if possible. That is what mature species do."

Deka paused. Kylac continued. "When we first met you, all you could think about was your job. If you missed too much work, you'd fall behind. As advanced as technology has made you, your basic everyday life is still a struggle to survive."

"That is what those people are doing," Deka continued. "They're doing what they want. If the portals were open, they would go to other words and experience life there, learn as much as they can about all the possibilities out there. Meet as many people as possible. Climb those mountains, run with those dinosaurs, experience life as a carnivore or an herbivore, talk to fish in the ocean."

Stephen didn't bother facing whoever was speaking. He kept his eyes on the sun—his sun, the little pinprick of light in the sky. "It looks like everybody's just lazing around or wandering, wasting time doing nothing."

Deka rubbed his claws. "You don't even recognize freedom when you see it."

"You know it's why fiction exists," said the fox.

"Fiction? What's that got to do with anything?"

"It exists because you are not free."

"Come again?"

"You are disempowered, both as individuals and as a species. Your lives are little more than a basic struggle to survive, and you are too busy surviving to go off on adventures in real life, so you experience it through characters who are doing the things you wish you could. Entertainment would cease to exist if people were free to do as they pleased and did not need to devote all their time to staying alive."

Deka's voice: "Instead of watching someone else have those adventures, you can have them. You can go to distant planets and experience what you once only saw in movies and read in books."

"It's how mature species live," Kylac resumed. "They don't need to be entertained because they can do all of those things themselves. They are free to develop their minds, pursue anything that interests them, experience life any way they choose. Real life can be as interesting as fiction if you are truly free."

"Your kind has been trying to liberate itself from the burden of survival for thousands of years. The need to find food, water, shelter, heat. It consumes the lives of animals, and every species has its own way of handling it. The solution your kind came up with is simple: shift that burden to another human being. The dominant males in your society have been using other humans for this purpose since before you were even sentient. Monarchies, dictatorships, religion, economics. They all have the same purpose."

"Dominant males?" said echoed.

"Your species is the right type to have such a social structure," said Kylac. "Before you became intelligent, one person in the group rose up to keep the others down in order to control reproduction. The less aggressive ones did as he demanded because they can't survive alone, and they can't fight him. You don't realize people still behave this way because you are inside of it."

"How do you know all this? You were only on Earth a week!"

"We told you, Stephen. Lone species develop according to their base instinct. As the species ages, its animal nature becomes more and more exaggerated. With no outside influence, it's the only possible way it can grow."

"This is what we do," said Deka. "Portals provide true freedom from all of that. We give people the freedom to live."

Stephen shook his head. "Is it really that simple?"

"This planet is ideal for finding perspective again," said the raptor.

They sat silently, watching the universe.

Loam

Stephen was the last to jump through the portal, and when he emerged next to Deka and Kylac, they were facing a dull red star. It appeared to be missing a chunk from it, and its glow seemed fainter than it should have been. Matter swirled around it, falling into itself, trying to coalesce.

Far in the distance of this dry world, the horizon fell off. Chunks of the land were missing, and enormous rods of soil, mantle, and planetary core floated in the sky. The atmosphere bled into space, making the wind strong.

"Guys..."

Deka answered. "I know this world. I can feel how gravity has changed. It's become uneven. I calculate more than half of this planet's mass has been lost."

"And exactly twenty-one point eight six percent of the star's mass is gone," continued the fox.

Stephen couldn't close his mouth as he stared at it. "Holy. Shit."

Bipedal rodents rushed around, grabbing unconscious reptiles from the ground and carrying them through an off-world portal in the distance. Others ventured out onto the surface to find more survivors.

A rodent with red skin covered in green fur hopped through the portal and noticed them. The wind blew the rodent's scent away from them, but they recognized the fur pattern and gait.

"Coe!" Kylac shouted.

Coe ran against the wind toward them. "Kylac, Deka! It's good to meet Archeons again."

Kylac dropped to all fours and ran to him. Deka tucked his arms in, streamlining his body, and sped after his fox, easily dashing past him and meeting up with the giant rodent first. Coe rubbed his muzzle against Deka's, and then did the same the with fox. Stephen jogged to catch up, the wind at his back propelling him faster.

The air rushed past them as enormous columns of planetary material broke apart and crashed to the ground in the distance. The star in the sky wobbled, flickered, and flung stellar matter into space.

"I was trying to make contact with the rest of the universe again," Coe shouted to be heard over the wind. "I opened a way to Loam, and this is what I found!"

"How long?" Deka yelled.

"Two days! We've been pulling survivors from the planet. They were conscious then, but now the air is too thin."

"We may only have another day before it's too thin for us!" Kylac said. "What parts of the planet have you searched? What's still left of the world?"

"The entire first continent is missing! I've been sweeping the settlements on the second continent from west to east, but I can only make one way at a time. I'm not sure how long the star will last. It may burn out, or it may destroy itself."

"Which settlements are left?" Deka said. "We'll make the ways there."

"Thel through Brair. Make the ways from Rebus!"

"That will take too long! We'll make the ways from here!"

Coe's tail flicked as he turned and ran back through the offworld sphere. Stephen had just caught up to them, still gazing up at the star, mouth agog. From his scent, Stephen's mind could not decide if it wanted to be in fight or flight, or crawl back into the womb.

"Oh my God. Is this what you told me about?"

The star flung more matter into space. Another planetary column crashed to the ground in the distance, splashing magma everywhere.

Deka turned his muzzle up to the star as well. His killing claws rose. "Finally!"

2

Deka and Kylac opened one way each, one right next to the other. The fox leaped through his portal, Deka through the other.

Stephen stood staring at the portals. Last time he had gone through Kylac's sphere. This time he went through Deka's. He overheard this was the settlement the reptiles named Ebe. The horizon dropped off halfway through it, meaning the people here had suffered the disaster directly.

Stephen stood in silent awe. Rodents poured through the portal behind him. Some were like shrews, others like rats, others like mice, all about as tall as Stephen and surprisingly stocky. Stephen never thought he'd feel puny compared to a mouse. The rodents fanned out, scented the air and the ground, and started pulling rocks off the piles.

Deka had told him the reptile species of Loam had once made piles of rock to impress mates. This produced a species of reptile that walked on its hind legs, and had opposable thumbs, as they were better able to carry rocks and arrange them.

They still did it, but now they created their own settlements, where water and food were more convenient. All of their settlements were under these piles of rocks, and many had collapsed when the disaster ripped the planet apart.

Deka then told him a fast story about when he and Kylac first came to this planet as children. Deka had shown an interest in aquatic species, and Kylac did not have to convince him to come here. Deka had spent days in the water watching the eels, trying to swim with them. He could not swim, so they carried him all sorts of places around the globe, even through underwater portals that led to oceans on other worlds.

Since the oceans had drained out into space and evaporated, the rodents had been scouring the remaining landmasses for other areas the eels might still be. The possibility that they may have lost an entire aquatic species disturbed Deka almost as much as the events on Neben had.

The raptor stood at the pile, helping the rodents pull rocks down and carry them away. Stephen ran to the base, climbed to the top, and began pulling rocks away. There was so much rubble, and it was spread out so far it seemed impossible they would find anybody alive under it.

The rodents were much stronger than Stephen was. The rocks were not very large, but where Stephen needed both hands and his legs to lift one, they lifted one rock in each hand. Deka also lifted rocks, but he had more muscle in his legs than some of these rodents had in their entire upper bodies. Even in the reduced gravity, Stephen felt weak. Fortunately, with several hundred people working together, it didn't take long to uncover the bottom.

Reptiles lay on the bare ground. Most of them had been crushed to death, but some were still breathing. Stephen gently picked one of them up. It was about the size of a komodo dragon, close to Stephen's height, loose skin of mostly browns and reds hiding a muscular frame, but still less muscular than the rodents. It was barely moving, but it flicked its tongue at Stephen and gazed at him, making croaking sounds. Stephen wished he could speak to it.

He carried it to the portal, then through a second sphere to Rebus. He set the lizard on the ground, relieved to be under a peaceful star and standing on a stable planet.

Various rodents were running around the field, tending to the wounded as best they could. One of the rat-like creatures with white fur and bright red skin pushed past Stephen and examined the lizard. He scented the lizard from head to tail, made some squeaking sounds at it, and then turned and regarded Stephen. The human couldn't help but stare at his crotch, as he had the biggest balls Stephen had ever seen. The rat-thing made some noises at Stephen, then walked away and examined another lizard.

The rodents were a bizarre rainbow of colors, like something out of the old *Star Trek* series: take an earth species, make it a different color, and now it's an alien. Here it was true. Rats with green skin covered in orange fur. Mice with red skin covered in striped purple and black fur. Shrews with white skin and brown and pink fur. It looked surreal, psychedelic, and funny. He would have liked to get to know these people under better circumstances.

Stephen looked back at the new hub. Five portals, all leading to different parts of that devastated world. He had been lifting rocks and pulling people out of the rubble for hours, and he was sore and terrified. Every time he set foot on that planet, an overwhelming drive to get off consumed him. He didn't know how Deka and Kylac stayed so long, making ways to each settlement so quickly.

Many rodents walked past him, everyone carrying reptiles. Most of the wounded had broken limbs, smashed jaws, or missing tails. Jellylike blood dangled from their wounds in thick strands that did not seem to break. Stephen looked through the portal again. He did not want to go back, but he didn't want to lie down and rest now, so he clenched his teeth, stood up, and walked through.

He emerged on the devastated surface of Loam, took another portal, and instantly he was back on the settlement of Ebe. Deka and one of the mouse-things were walking one of the lizards up to the portal. Stephen stared. Until now, he had not seen any of the reptiles walk, and he was surprised they were bipedal. The mouse-thing had bright blue fur with orange spots. Tired as he was, the human managed to laugh.

Everyone was still pulling rocks away, still pulling people from the rubble and carrying them to the portal. Some of the injured reptiles could walk, and as they came toward Stephen, one of them caught his eye. It was carrying an armful of something, and after a few strides Stephen realized what they were.

The ground shook. Stephen held his arms out to keep his balance. The reptile walking could not do that, and she fell, the eggs she was carrying spilling from her arm and rolling everywhere. She frantically reached and tried to collect them

Stephen ran toward her and began gathering the ones that had rolled the farthest. They were leathery, twice as large as a chicken egg, but more cylindrical than circular. Stephen collected six for her. She caught up to him, carrying the other three, flicked her tongue at him several times. Stephen leaned to be closer to her eye level, and she felt him with her tongue.

After a moment, she turned and kept limping toward the portal. Stephen's first instinct was to pick her up and carry her through, but he wasn't sure if that was a good idea, since she looked sharp and dangerous, so he just walked with her.

The planet shook again. Stephen kept his footing, but he noticed the lizard was about to fall and held her up around the shoulders until the ground stopped moving.

Her tongue flicked over his hand many times, and they walked through the portal. The way directly in front of them led to Rebus, and they stepped through one at a time. Stephen wasn't sure where to go now, but she seemed to know. She limped away from the vast field of wounded, and Stephen followed her for a distance. She set her eggs down and lay next to them. One at a time, Stephen set the eggs he was carrying with them and backed away.

She made some throaty noises at him.

"You're welcome," he answered. "Good luck."

He turned away, ran back to the portal, and leaped through. Deka and Kylac were here at the temporary hub, standing in that way they did when they were making a new portal and it was taking a lot of concentration.

Kylac smelled him approaching and looked at him over his shoulder. "People are trying to figure out what the hell you are and what you're doing here, but you're helping, so they like you."

"I'm glad. By the way, is it rude to notice these rodents have humongous balls?"

Kylac's lips curled at the corners, and he smiled like a human. "They do, and they're not even the biggest out there!"

"They're not?"

Kylac smiled bigger, and now he laughed with his voice. "If we can, I'll take you to Jemum. Those people... I don't know how they live!"

"Kylac..." Deka grunted.

"I know. I haven't forgotten where we are."

A portal appeared in front of Deka. Moments later, another one appeared in front of Kylac. Both ways led somewhere devastated and terrifying. Each leaped through his own portal, leaving Stephen staring at them.

Stephen leaped through Kylac's sphere. The fox was already running toward a rock settlement, nose to the ground, scenting everything from here to there.

Stephen noticed the horizon had fallen away. He walked to it, carefully at first, in case of another quake, then faster and bolder. He perched on the ledge and looked down. It reminded him of those textbook diagrams of the interior of Earth, with a quarter of the planet cut away to show the different layers beneath the surface. He was standing on the edge, looking down that cutaway, the layers of the planet reaching far, far below. Many sedimentary layers followed by a layer of magma that had been leaking out for some time, forming globules that floated in the empty area where the planet had been. Much of the magma still glowed hot, and some of it hung in orbit around Loam.

Stephen looked up at the star. Though it was so far away, even from this distance he could see it rotated unevenly. It wasn't possible to see the sun spin with the naked eye, so seeing this star spin and wobble wildly made Stephen nervous. It was shedding mass, throwing so much of itself into space as it tried to form a sphere again.

The ground began to shake. Stephen backed away from the edge and ran. The ground shook harder, throwing Stephen down, and he waited until the world stopped moving. When it did, he brushed himself off and ran to the rock pile. Hundreds of rodents were pulling the rocks down by now. Stephen took a place and began casting the rocks aside.

Kylac nudged Stephen in the ribs. Kylac was snickering, mouth parted in a human grin, showing teeth and laughing vocally, pointing at a shrew's testicles, which were as large as Stephen's clenched fists.

Stephen tried to hold it in, and then giggled with Kylac. He felt ten years old, but it felt good to laugh with someone again. He hadn't realized how long it had been since he had.

3

Deka and Kylac stood on their hub on Loam, preparing two more ways to two different settlements. A hundred rodents and about twenty able-bodied lizards stood by, ready and waiting.

Deka turned and faced Stephen. "I'm making a way to Thess, a settlement down in one of the valleys. It will be warmer there than where Kylac is going."

Several tongues touched Stephen on the back, on the shoulders, on his rear. All twenty lizards were behind him, taking in his scent, trying to figure out what he was. He was used to it by now, and it didn't even tickle anymore.

Deka rubbed his claws together as he eyed him with that immobile face of his. "They like your scent. It's exotic. If one of them wraps his or her tongue around your nose, they're courting you. Remember to be a good ambassador." Deka lightly clawed Stephen down the arm.

Stephen smiled, rubbed his arm where Deka had scratched it. "Tell them no thanks. After seeing all this, I don't think I'll ever be in the mood again."

Kylac elbowed him on the side, grinning ear to ear. "If you want, I'll ask one of the Rebens to let you feel his balls."

Stephen laughed. Kylac laughed with him.

Two portals opened next to one another. Deka leaped through his sphere. Kylac leaped through his own. Stephen followed the fox.

It was chilly here, but Stephen took one look at the rock pile in the distance and felt sure he was about to work up to a sweat. The rodents ran around them, climbed up the rubble, and began moving rocks down it. Others picked up komodo-sized reptiles where they had fallen and carried them through the portal.

After an hour of this, the weak gravity finally caught up to Stephen and pulled him to his rear. He sat, panting and sweating. All this heavy lifting, and it had been years since he had last done PT in the Army.

"Is there any water?" he said to Kylac.

Kylac shook his head. "I wouldn't drink the water here. Go back to Rebus. Theirs will be clean."

"I need a drink. Oh, God, I'm tired. How many more places are there?"

"Twenty-six more." The fox glanced at him. Stephen swore he saw an expression of dread on his face.

"Will we finish in time?"

"Hopefully now that three Archeons are working together."

"God, I hope so. I'll be back."

Stephen climbed down the rock pile and began walking to the portal. Several reptiles were hobbling toward it. One of them was having a difficult time walking. Stephen noticed the sticky blood clinging to one leg, and a bone jutting out of a large wound. Stephen changed course and walked beside it, offering support. The creature didn't seem to know what Stephen was doing. It hobbled a few more steps, thick blood squeezing out of the compound fracture.

Stephen wrapped an arm around its shoulders. Now it held Stephen and took weight off its leg and leaned on him. It must have weighed more than Stephen, despite being a foot shorter, as the stress on his back and legs felt immense. The lizard never stopped hissing. Stephen walked it through the portal and then to the refugee camp on Rebus. He found an empty place and set the thing down. It rested calmly on the ground, flicking its tongue across Stephen's face. Stephen was glad to let go. Too sharp for his tastes.

He looked around, found the lake. Many lizards and rodents had gathered around it, lapping water from its calm surface. Stephen walked around the wounded and the rodents and lizards tending to them. There was a lot of wound-licking, and a lot of snapping as people reset bones. Their hisses of pain did not make him wince.

Stephen walked by a few mice. The females were hard to ignore, since when they bent over to tend the wounded, everything was visible. He wondered how anyone concentrated with that staring them in the face all the time.

The lizards, however, had nothing visible outside, and he couldn't tell them apart, male, female, or individual. They looked the same, acted the same, moved the same. At least the rodents had different fur patterns.

He reached the pond and knelt in front of it, cupped his hands, and brought a handful of water up to his face. He did this several times, and then noticed everyone staring at him.

Stephen waved at them, cupped his hands, and took another drink. The lizards and rodents resumed lapping the water from the pond.

4

Deka and Coe stood before their hub on Loam. Stephen waited among the rodents. He felt snouts sniffing him, but Stephen barely noticed them anymore. The Archeons opened one portal each and leaped through. Stephen followed Coe. He wasn't sure where Kylac had gone, but he wasn't worried about him.

There was no rock pile here, which surprised Stephen. He had been moving rocks for so long he thought there was nothing else to this planet. Stephen ran to catch up to Coe and then kept pace with the green rat.

"Where is it?"

"I am not sure," the rat answered, in English.

They and the entire throng of rodents and lizards ran to the ledge. Some stopped, picked up various unconscious reptiles off the ground, and carried them back to the portal. Others stopped at the ledge and looked out across it. The section of land had broken off into pieces and was now floating some distance from the planet.

Stephen turned to Coe. He was concentrating on that floating island.

"We're not going there, are we?"

"There may be survivors."

"What about air? It's bound to be gone by now."

"I'll add a spin to the portal here. It will draw in just enough air for us to breathe."

Stephen glared at that floating island. His heart raced. While Coe stood still and calculated a way, others scurried around, grabbed some people, and carried them.

A sphere opened. Air rushed past him and fell into it. Stephen waited until Coe walked through, then he followed. Stephen felt as light as a piece of bread.

The chunk of rock floated far over the surface of the planet, tilted at a high angle. The rocks on its surface didn't rest on it, but drifted just above it. Coe was hopping along, taking long leaps. He stopped, shoved and kicked rocks away, and reached some survivors on the bottom of the pile.

Stephen took a running start and fell upwards. He waved his arms, expecting to fall to the ground, but he wasn't falling—he rose up and up and up and up. The air became thinner and colder. Stephen held his breath. Goosebumps rose all over his body. His ears popped.

Below him, Coe and several rodents scampered along the surface as if they'd spent their whole lives tossing unconscious reptiles across a floating chunk of planet. One of the rodents noticed Stephen and flicked its tongue in his direction.

Finally Stephen reached a zenith, and he slowly fell back down. The rodent waited until the human had fallen far enough, then grabbed his leg and pulled him the rest of the way. Stephen stayed firmly planted on the surface. He wanted to run and help but this was definitely not the place for him, so he retreated through the portal to a surface with stronger gravity.

He knelt, shivering and gasping for air, rubbing his arms as rodents began streaming through the portal, many carrying wounded. One of the reptiles walked up to him, flicked its tongue over him, made some hissing and throaty sounds in his direction.

"Hi..." Stephen said.

The reptile flicked its tongue over his face, moving close. Its scales were warm, but felt flimsy and gross. Stephen shied away. The reptile seemed to understand and limped back through the refugees and the rodents tending to them.

5

The unstable sun in the sky had set and risen twice. He hadn't seen anyone sleep since he first arrived on this broken world, so Stephen had remained awake this whole time. He had no idea how long a day was here, but he was starting to feel like it was way past bedtime.

Stephen emerged at the back of the crowd flowing into two new ways to two new sections of Loam. The reptiles poured into the portal on the left, while the rodents filed into the sphere on the right. Stephen followed the rodents and landed in another place that lacked a rock pile. Everyone was picking up reptiles where they lay and carrying them back to the sphere.

"There you are," said a reptile voice behind him. Stephen turned and saw Deka walking up behind him. "Heard you almost launched yourself into orbit."

Stephen laughed dryly, rubbing the back of his neck.

"Everybody has been to worlds with weaker gravity," Deka continued, holding his claws together, his face as stiff and rigid as ever. "They know how to move in those conditions."

Stephen backed away a step. He never noticed it before, but Deka's face always looked like it was hunting him. Stephen covered his nerves with a laugh. "My first time... I almost killed myself."

Deka rubbed his claws together and then reached out and lightly scraped Stephen down the arm. It left raised red marks on his arm. Stephen took a half-step back.

"You were never in danger. Coe could have made a sphere in five minutes to send you back to the surface. Over a distance that short, he could have made a flat portal fifty feet square to catch you as you fell. Kylac and I were going to take you somewhere to practice moving in low gravity, but plans changed. Be aware. There are things here that can kill you."

Stephen grinned. "Now you tell me."

"It should be obvious, so I did not think to state it. Please don't get killed. It will be difficult to explain to your sister."

Deka was rubbing his claws together, so he was laughing, but it didn't feel like a sardonic joke. Stephen felt threatened.

"Yeah..."

Stephen felt footsteps coming up from behind. He turned and saw Coe.

"Deka, we're clearing up here. You can make another way now."

"I will."

He turned around, walked back through the portal, stood on the stable part of Loam, and concentrated. Other large reptiles gathered around the raptor, tasting his scent with their tongues, crowding him. Deka rubbed muzzles with them.

Stephen looked at Coe, who was walking away. He followed the rodent and searched for more survivors.

6

Stephen had heard it was the last pair of settlements, and given the way the star in the sky thrashed, they weren't a moment too soon. The air here had become much too thin for Stephen's comfort, and it was also becoming too cold.

The three Archeons were talking about Stephen's first encounter with weak gravity. It struck Stephen as odd they should be doing so in English.

"Stephen almost launched himself?" Kylac giggled.

Stephen smiled back. Seeing someone else laugh helped him feel better about it.

Kylac smacked Stephen on the back. "We definitely have to teach you how to survive better!"

Deka stood on his other side, rubbing his claws together. He reached out and raked Stephen's arm again. The raptor never drew blood when he did this, but Stephen's arm was raw by now. He leaned away from him.

The last two ways opened. Deka ran into his sphere. Kylac and Coe ran through the other, the fox still giggling. Stephen followed the people with fur.

This time the rock pile before them had not collapsed completely, and the others ran toward the entrance. Stephen did not want to enter that thing, but he trusted everyone knew a safe mound of rocks from an unsafe one. He followed them inside.

The entrance was large, and as Stephen's eyes adjusted to the dark, he began to make out the structure. It was completely open, only about ten feet high, and stretched far enough to vanish into darkness. It was just rocks piled on top of one another, frequent columns interrupting the internal space, also made of loose rocks. It looked unstable, but somehow the stones held one another up. Lizards rushed by, rodents ran around him, searching for survivors.

The rocks had been painted from floor to ceiling. Beautiful, intricate designs everywhere. Stephen paused and examined one of the designs by the entrance, ran his hand down it. It was made of dried skin. He followed more designs up the wall and across the ceiling. The entire structure was decorated with their own shed skins.

Moments later, the group of lizards and rodents came rushing back out, more reptiles following. He smelled a canine standing next to him.

"Tell me, Stephen," said Kylac. "Why did you follow me and not Deka?"

"Well, I had to pick one."

Kylac laughed. A human laugh. "Deka and I noticed it after the first two portals. You're more comfortable around me than with him, and with the Rebens than with the Loam reptiles."

Stephen shrugged. "I have to pick one, so I just pick theirs."

Kylac did not smile. His tail flicked from side to side. Suddenly he seemed distant and alien again. "It became more apparent when I started to imitate your facial expressions and laughter."

"Yeah, I noticed."

"I just stopped. Feel a difference?"

"Yeah... yeah, you're... You're not as... Not as..."

"Not as friendly as before," Kylac finished. "That's what you're feeling. But I am still canine, so you still feel more comfortable with me than with any of the Loam."

"I suppose, yeah. They're sharp. Their tongues are... kinda gross."

The fox's tail waved again. "And you shied away when Deka joked around with you."

"It's probably because you're cuddly mice and dogs, and he's a flesh-tearing dinosaur."

"No, Stephen, that's not why."

His face was still again, but his tail was wagging. Smiling. It took Stephen conscious effort to remind himself of that.

"You're drawn to what's familiar. Deka's way of smiling and laughing is alien to you. The Loam relate to each other in very strange ways, too. The Rebens, however, are more familiar. They have fur, you can tell the genders apart, they sniff you. You know what all of that means. Since I started laughing and smiling the way you do, you're choosing me over any of them."

Stephen didn't know if he should confess or not. Kylac continued.

"Deka feels cold and emotionless, doesn't he?"

"Well, yeah, a little. He doesn't express himself much."

"He does, but your mind is latching on to what you already know. Mature species learn to seek the unfamiliar, Stephen."

"Uh. Sure."

Kylac wrapped his arm around Stephen and began walking him deeper into the structure. "You seem a little put off by how they decorate their homes. Taking their own skin, cutting it up into patterns, using it as a canvas. Do you know how they attach it?"

"No, and I don't really want to."

"Think about it from their perspective, Stephen. They shed their skins once a year. That is also the only time they sleep. Everyone else in the community strips their skin off while they sleep. Think about that. You go an entire year without sleep, then you wake up, and you're a new person, both in body and mind. All the information they took in over the previous year settles, and it changes them. Their old skins represent the person they used to be, one with less knowledge, less maturity. They commemorate their old selves in this way."

Stephen looked up and around at the structure. Sheets of dried, shed scales of various designs hung everywhere.

"It sounds kinda morbid. Like decorating the baby room with grandma's skin."

"Embrace the unfamiliar, Stephen. Humans will not be able to explore the universe until they learn how to do that. Stop regarding the strange as a threat. Stop being drawn to what you already know. If you had a companion species, it would not be your first reaction. You would want to learn more about them, experience life as they do."

The ground shook. A few of the rocks from the ceiling came loose, falling near Stephen and Kylac. Two of the skins also detached and drifted to the ground. Kylac walked to one of them and picked it up. He scented it, stood up and handed it to Stephen. The human hesitated, but then reached out and took it. It felt light as tissue paper, and now he saw exactly how it had been attached to the rocks. The back was covered in blood.

Stephen dropped it, wiped his hands on his pants, and then remembered he wasn't wearing clothes, so he had smeared his legs in Loam blood.

"Their blood doesn't dry as yours does, Stephen. It remains sticky. Old survival mechanism; they left a trail of blood, and their predators would step in it and slow down, giving the Loam reptiles time to run away. It also seals wounds and traps infection. Now they use it for this."

"That is... weird."

"Says the human who drinks breast milk that isn't from a human." His tail waved, and his ears pointed forward, but his face remained immobile.

Stephen smiled back.

The ground shook again. More rocks fell. Kylac took Stephen's hand and ran for the exit as the tremor increased. As soon as they were out, the human and the fox dropped to the ground and held on as the entire planet seemed to shift beneath them.

The structure behind them wobbled and crumbled as the individual stones gave way and cascaded down. Stephen flipped over, propped himself up, and watched as the whole thing caved in, kicking up a giant cloud of dust. Finally the ground settled.

Stephen panted. Shook his head. "My God... I still can't believe I'm seeing this. I'm seeing it... It's not a movie. This is the real thing."

Kylac was on all fours, scenting the ground, dashing to the broken edge of the planet. Stephen scrambled to his feet and ran after him.

The ground heaved sideways. Stephen struggled to stay standing, but Kylac kept running.

"Kylac, what are you doing?"

The planet shifted harder, and Stephen looked at the broken horizon. Ground was falling off it, the far end of the structure slipping below the horizon, and still Kylac was running closer.

"Kylac! Come back!"

The fox was on all fours running headlong toward the falling ground. The planet thrashed so hard Stephen couldn't stand. He had no idea how Kylac was doing it and taking in scents at the same time. The fox paced back and forth, circling around and around, nose never leaving the ground as the horizon came closer and closer.

"Kylac!"

Stephen couldn't hear himself over the roar of collapsing rock. He turned to the structure. Half of it was gone, and the rest was gradually falling off the horizon. Pieces that had fallen away were now in orbit far in the distance. The rumble increased, and the planet shook harder. Stephen tried to stand, but all he could do was grip the grass-like plants with all his might.

Kylac stopped suddenly and buried his nose in that one spot for several seconds. Finally the fox seemed to become aware of where he was and ran back toward Stephen, the horizon on his heels. Even from the ground Stephen could see pieces falling away, rising into orbit and floating gently in the distance. Kylac ran ahead of it, making far leaps, staying just a few feet ahead of the crumbling edge.

At last the ground held still. Kylac kept running all the way up to Stephen and then slid to a stop in front of him, panting. He held a paw out to Stephen and helped him up. As Kylac rose to his hind legs, they turned to face the new horizon. It was only a few dozen feet away now.

The fox took off running toward the portal. Stephen ran with him, trying to keep up but falling far behind.

"What's wrong?" Stephen called.

"I have Friend's scent! I know where they went!"

7

A lone portal to Loam showed its daytime star in the sky. The star had finally collected itself again, and the sudden force of all that matter coming back together forced a fusion reaction. The star expanded, doubling in size in mere breaths, and tore itself apart, scorching Loam.

The hundreds of lizards watching through the portal mourned their world, but at least their companion species had not been lost. Even as they grieved, some made their way to the lakes and touched a clawed hand to the heads of the eels. Vibrations transferred though the skin, and they told the fish what was happening.

Some of the Rebens had found pockets of water still on the planet and had carried the eels off. Only a few dozen had survived, and the water did not have as much oxygen as they were used to, but they were alive. The Loam race was complete.

Coe allowed the portal to close, and then he walked to the lake. The Relian canine and the human sat on the shore. Deka stood in the water, communing with the eels, taking in their grief and comforting them as best he could.

Four reptiles approached Stephen, tasting his skin as he raised a handful of water to his mouth. Stephen looked at the one reptile in the foreground. He or she looked like the walking dead, with scales peeling off, eyes cloudy and also peeling. Deka noticed and raised his snout from the water.

"Do you recognize her?"

"No."

"She says you helped her carry her eggs."

"That's her? What happened to her?"

"She's about to sleep," Kylac said, tail waving wildly. "You carried her eggs. That's a courtship gesture in their culture."

"What?!"

Deka spoke now, releasing two of the eels to hold his claws together. "The male who fertilizes the eggs, and the male who helps raise the eggs, is never the same person. To carry someone's eggs for her is a proposal."

She was still staring at him. Stephen turned around and sat down, meeting her eyes, leaning as far away as he could.

Kylac grinned like a human. "Remember how you started to feel comfortable around me when I laughed with my voice? You did something she understood, and now she feels an attachment to you."

Stephen couldn't blink as this komodo dragon-like thing stared him down. "Shit. I don't think I'm ready for this."

"Oh, don't worry," Kylac said, grin widening and tail waving. "She's not stupid. She knows you didn't know what it meant. But the emotion is still there."

"She's only asking you to help her sleep," said Deka as he dipped lower into the water.

"What's that mean?"

"Take her skin off."

"Take her—?"

The lizard in front of him could barely stand. She closed her eyes and dropped to the ground in front of Stephen. Stephen stared at her. He started to get up, then met Kylac's eyes and froze. He looked from Kylac to Deka, to the other lizards around him, to the rats. Finally he sat back down again and looked at her.

"What do I do?"

Kylac and Deka made some noises to all the reptiles in the area, and they started flocking to Stephen. The human's eyes widened at the sight of so many sharp claws and rough scales and flicking tongues coming at him.

"They'll show you," Deka said as he rubbed one of the eels.

Stephen became lost in a sea of lizards, and Kylac allowed his lips to rise in a smile.

Coe approached from behind Kylac and stood facing the Relians. "I've been told you have news for me?"

"Yes," Kylac answered. "Stephen wants to feel your testicles as soon as he's done."

The muscular rat twitched his ears, looked sideways at the human. "He stares at them enough. I was waiting for him to ask."

Kylac's tail waved. "He doesn't know he wants to though."

Deka wanted to rub his claws together, but he was too busy trying to touch all the eels swimming around himself.

The fox continued. "I found Friend's scent at the last settlement."

"Did you learn anything?" said Coe.

"I followed it to where they made a way offworld. Rive let the portal intersect the ground on both sides. I smelled where they went."

Deka growled. Some of the eels wrapped themselves around him, sensing his distress and trying to comfort him. "At last we know where we're going."

The only sounds came from the mob of reptiles showing Stephen how to help a Loam lizard sleep.

Uiv

The sphere did not close immediately behind them this time. The star in the sky appeared to be whole and stable. The horizon was distant and smooth, but the ground itself was made of loose rocks and rubble, as if a cliff face had collapsed and filled the land with jagged debris.

"Nothing's wrong here," said Deka.

Kylac was on all fours, scenting the rocky ground. "They were here! Definitely! I can smell Friend, and Rive's foot!"

"So either they're still here, or they left without destroying this world." Deka looked back at the portal he opened. "I'm trying to keep it open. It's already starting to hurt. I'd feel a lot safer if I could keep ways open longer."

Kylac had wandered far off, following the scent. Deka chased him. Stephen carefully stepped over the jagged rocks and caught up. He hadn't even bothered to put clothes on this time, but he wished he could pause to slip on his shoes.

"Guys! Uh, how long's it been since we left Earth?"

Deka heard him, but he was too far away to answer now. Kylac was running along the scent path headlong into the planet's only settlement, just over the next hill. Deka bent down and scented the ground. He smelled Friend as well, but he had to scent it several times to build up enough for him to detect. It was much stronger for Kylac, much easier to follow.

Kylac crested the hill and paused. Deka dashed up the hill and stood next to him. Stephen trudged his way up and stood next to the Archeons, looking down at the settlement below.

A gigantic tower made of soil reached high into the sky. Something hard and shiny held everything together, giving it a reflective sheen. There were no sounds anywhere. Even the wind seemed calm here.

"The scent doesn't go that way," Kylac said. "Fill Stephen in. I'll follow it."

Kylac dashed off in the opposite direction. Stephen turned and watched him.

"He is so much like a fox. An Earth fox, I mean, when he runs like that. Deka, did you hear me?"

"I did. You've been off Earth for about two months."

"Two *months*... Uh, is there anywhere I can take a bath? I haven't washed since we left and I'm covered in dirt, sweat... Reptile blood. I haven't even brushed my teeth. I'm filthy."

"Stephen, listen closely. Meet my eyes, please."

Stephen looked at the raptor. Deka lowered his head to the human's eye level.

"The car ride is over."

"It is?"

"We're on Friend's trail, and we have to follow it. What you saw on Loam is what he has done to at least five worlds, and it will continue until someone stops him. We will travel to planets that will not be safe for you. It has only been two months, so you should be able to slip into your life on Earth again with ease."

Stephen took a few breaths.

Deka raised his head to a comfortable height. "You've behaved yourself very well over these weeks. Kylac and I are both impressed by your versatility. You adapt very well to unusual circumstances, but what you've experienced only scratches the surface. This place will be your last taste of what is beyond your world. Take these people, for example."

Deka turned to the shiny tower of dirt and began walking down the hill. Stephen followed close behind. The loose rocks gave way to soil at the bottom.

"What you're about to see will test you. The people you're about to meet are called the Uiv."

"Oo-ihv?" Stephen said, stumbling over the odd syllables. "Who is their companion?"

"Both species are known as the Uiv because you can't separate them. They are essentially one species."

They walked, but the tower did not seem to come any closer. Stephen shrugged his backpack and kept following.

"That thing looks as big as a city," said Stephen. "Jeez, it's all the way up to the sky. I thought you said normal species don't build things."

"For some, it is their nature to do so. I told you about Hithe."

"You said it's not normal, but it is for them?"

"The definition of normal depends on the reason people build things."

"Uh huh. So what are they? More lizards?"

"One species is a mammal. Feline. Their companion species is a type of insect."

"An insect? That'll be different."

"I'd take you to Kattaaka, but that world isn't safe for anyone. Everything is venomous, and now they have a new companion race to understand. A former predator. Anyway, Stephen, there won't be time for introductions here, so you must be ready to see something distur—"

Deka stumbled. Pain pierced his skull, and he dropped to his knees, screeching and growling. Stephen knelt beside him, holding his shoulder as another wave of pain shot through Deka's head.

"What's wrong?"

Deka was panting through clenched teeth. The off-world portal was slipping away. He was trying to hang onto it, but it wasn't happening automatically. He concentrated, tried to hold it—then it fled.

Deka snarled and screeched, toe claws digging the dirt. Stephen backed away as Deka slowly rose to his feet, claws still raised. He turned to Stephen. Deka noted the human's pounding heart and tried to lower his claws to appear less threatening.

"The way back to Rebus. I lost it. I was trying to keep it open until we knew where Friend was. I can't even do that. The disaster took that away from me."

"But you can make a new way any time, can't you?"

Deka winced at the lingering pain behind his eyes. "The planet has already moved millions of paces away. To make a new way, I'll have to catch up to it. I had it. It slipped away from me." He snarled at the ground. "It never used to do this! I hate being like this! I'm not an Archeon anymore!"

Stephen stood frozen. "I'm sorry. I can't even imagine what that's like."

Deka huffed and began walking to the tower. Stephen followed from a wide distance. A quadruped stood a few dozen paces in front of them, vaguely feline in shape, no fur, and her skin was red and covered in raised blood vessels.

Deka switched to the native language. "Greetings."

She turned to look at them, her muzzle dripping blood. Stephen stopped and glanced at Deka, who did not slow down. Stephen jogged to catch up, and then walked at his side. The hairless feline was as high as Stephen's chest at the shoulders. An insect as large as the cat's head lay on her skull.

"Greet-" said the feline's muzzle.

"-ings," said the insect.

Stephen stopped again. This time he did not catch up. Deka walked until he was in scenting distance and then held his hand out to her. The muzzle scented him, then Deka raised it to the insect on top of her skull. It extended a thick tube from its mouth and tasted his skin.

"Stephen," Deka said, in English. "Introduce yourself to the Uiv."

"Deka, why are that bug's legs inside the cat's skull?"

"I'll explain later. Please come here and do as I just did."

Stephen carefully walked closer and stood next to Deka. The feline/insect turned and faced Stephen. Stephen held his hand out. The feline scented it. Stephen did not raise it to the insect's level. Deka reached out, grabbed his hand, and raised it for him. The insect tasted Stephen's skin and felt it with its antennae.

"Deka, this is gross. What's it doing?"

"It's making sure we're not compatible."

"What?"

"Mentally. It's actually considered very serious for a Host to be without a Guest."

"Host? Guest?"

The insect retracted its appendage. Both it and the feline now stared at the human.

"The insect is a parasite. The feline is its host."

Stephen glared at Deka, then glared at the furless cat with the insect attached to its skull. The blood vessels on the top of the skin looked as though they would burst at any moment. Deka scented Stephen discreetly. As he expected, Stephen was on the verge of panicking.

2

Kylac followed the scent trail over the desolate land. Much of the land was bare dirt and loose rocks dotted with little stagnant ponds full of biting insects. He was climbing over the rubble pile that formed much of this region. Loose, shifting rocks ranging in size from bigger than his head to as small as his claw. Nothing grew here, as the soil was sterile. Even after centuries, the region had not recovered from the attempt at genocide.

The scent was difficult to find here, but Relian canines were finely adapted to scenting out their own kind, and Friend's stood out as the only fox scent he had caught in years.

Rive and Friend had walked far from civilization. He could smell the frustration and anger in Friend's scent. He could even catch something strange in Rive's, weak by comparison because it was only a single foot. Some of the rocks shifted under him, but Kylac had sure footing on all fours.

It took Kylac several dozen more paces before he collected enough of Rive's scent to figure out what was so strange about it. Rive was afraid. Kylac paused and thought about all the things Rive would be afraid of: his fox destroying entire worlds. Being unable to open an offworld sphere in time to escape the next antisphere. His own body.

Rive was some sort of metallic machine now, rebuilt by an uncontacted species that had since been destroyed. He wondered how that must feel, to be a machine. Kylac could not imagine it, and he wondered about the people who could create a technique to merge flesh and metal. They must have been a troubled culture, acting out their collective loneliness by, in essence, creating their own companion species. Kylac did not blame him for being afraid. Kylac would be afraid, too, if his body now relied on machines he did not design, could not repair, and did not understand.

He picked up the scent trail again, following it down the aggregate slopes of rubble. He wound around ponds. Small insects rose from them, flying at his face and landing on him, but they did not recognize his scent, so they did not bite. Friend's scent seemed stronger now. At the same time, Rive's scent became more agitated.

Kylac caught a new odor, one he did not expect to smell this far away from civilization. He raised his head and saw it. An Uiv was walking across the rubble toward him. Kylac took in its scent, determined it was male.

"Greetings," Kylac said in the Uiv language.

"Hello," said the feline.

"Hello," said the insect attached to his skull in exactly the same tone and cadence.

"Why are you way out here? I didn't expect to meet anybody this far."

"Probably for the same reason you're alone," said the furless feline. The blood vessels on his body were so red and raised he appeared to be bleeding.

"Are you tracking another Relian canine Archeon who is destroying planets by thinking about what exists outside the universe?"

"No."

"Then let's hear your reason," Kylac said, waving his tail around, sitting as a quadruped.

The Uiv sat down. Insects from the lake landed on him, sank their jaws into him. Kylac was about to say something, but he held back. The feline stared at Kylac in silence for a long time. He smelled as if he were trying to find words. Finally the insect on his skull spoke.

"He is trying to kill me."

Kylac's ears bloomed.

3

Hairless lions huddled together like insects in a colony, and they barely left any room for Deka and Stephen to move. The tower base was large enough to be a city, and this was only the first layer.

"Now, Stephen, nothing here will hurt you, but I won't leave you alone to explore this time."

"Please don't. This is..." Stephen shivered instead of finishing the sentence.

Deka led Stephen straight through the middle. Feline/insects were standing and sitting, talking. Some were eating prey, a vaguely equine species that still twitched as they tore into it.

Many caught Stephen's scent, stopped what they were doing, and approached. Stephen let them come near, but Deka noticed his eyes never settled on the insects. Deka leaned close to the human.

"Refusing to look at something does not make it go away, Stephen."

"How can I look at that? It's—"

"You handled the Heu very well, but I can tell this race hits a tender spot. Why do you find it repulsive?"

The Uiv had surrounded him. They were asking him what he was, where he came from, if he was an Archeon, and so on. The feline said half of each word, and the insect spoke the second half.

"There's a bug attached to their brains! Its feet are in its skull!"

"I promise they will not do anything to harm you. You're not the insect's type. Nobody but the Uiv is."

One of the felines spoke in English. "Hel-lo-Ste-phen-what-are-you-and-is-this-your-first-time-off-world?"

Stephen watched the Uiv as each syllable came from a different mouth, forming a seamless sentence of two voices. Stephen tried to stay composed.

"You-smell-af-raid," said the Uiv. "Are-you-all-right?"

Stephen did not answer. Deka spoke up.

"Stephen, this is Ogu, one of Uiv's Archeons. We met her at the entrance, remember?"

"Yeah, yeah," Stephen said. "Hi. Ogu..."

"Your-lan-guage-is-nice," they said. The insect embedded in her skull had spoken the first word. "I-can-tell-you-are-a-vis-u-al-spe-cies."

Stephen had frozen up. Deka reached over and lightly clawed him down the arm. It shocked Stephen out of his stupor.

"Sorry!" he said. "I've... I've never seen anything like you before. I don't know what to make of this place." Several others bumped into Stephen, pushing him still closer to Ogu. "It's cramped in here. What is everyone doing?"

"They-are-wond-er-ing.   They-are-learn-ing-a-bout-the-uni-ver-se-phy-sics-math. Since-the-port-als-are-closed-they-have-to-learn-with-out-ex-peri-en-cing.   Ve-ry-time-con-sum-ing."

Deka noticed Stephen was barely listening, so he stepped in. "Ogu, Stephen would like to know about your people. Where he comes from, insects that bury their legs into someone's skull are a bad thing, so the human is feeling a little culture shock."

The insect on her head clicked. The feline huffed. Deka just now realized how close it was to human laughter. He hoped it would help. Normally this was when he would leave Stephen alone to experience the people for himself, but Deka wanted to be nearby in case the human panicked.

4

Kylac remained on all fours as he followed the feline to the tiny settlement. It was little more than a depression in the ground with most of the rocks cleared. Various felines sat with heads hanging down, staring at their paws. Others lay curled in on themselves, trying to sleep. They smelled as if they hadn't slept in days.

"Who is that?" said one of the felines.

"This is Kylac," answered the one leading the canine into the depression.

Everyone in the bowl snapped to consciousness and rose to their feet, twenty-seven of them in all, and Kylac did not like their collective scent. Something about them was not right, and Kylac had a good feeling he knew what it was. The only reason he allowed the Uiv to lead him here was because Friend's scent trail came through this place. Now that he was in here, the trail was gone, probably worn away by all the people walking about.

They approached and scented him.

"Someone like you," said one of the felines. The insect atop his head had no part in the words.

"I heard about Rel. Have you found any survivors?"

More questions hit Kylac. They came closer, and Kylac backed off. Everything in him told him to keep them away, but they closed in. Finally Kylac spoke up.

"Everyone stay where you are!"

They did, still smelling of agitation and fatigue.

"I'll answer all of your questions if you answer mine first. Did another fox come through here?"

One of the insects answered. "Yes. Days ago two Relians were here. The theropod was... I do not know how to describe him. The canine smelled strange, too." The feline it was attached to had been silent the whole time, which looked and sounded eerie.

"Did they speak to anybody? Did you speak to them?"

"I tried," said one of the cats. The insect atop her head did not speak. "But they kept walking."

"Did they say anything to you?"

"Nothing," said one of the felines. Then the insect attached to her skull spoke. "We tried to speak to them, but they did not answer. They did not visit us."

"Where did they go?"

"To our lake," said one of the insects.

"Thanks. I'm following their scents. I will return."

Kylac walked deeper into the depression. Friend's scent was gone, so he walked in a straight line all the way to the water. There he saw a circle of discolored soil at the shore. He approached it and scented it deeply for some time, making sure of where it led.

He looked over his shoulder. All twenty-seven hairless felines were behind him, glaring at him. Their insects seemed to stare at him separately, and they smelled even more off than usual.

"Please tell me," said one of the felines. "What's it like to be alone?"

Kylac stared. Everyone stared back. The fox turned all the way around. "You could starve out here. Why are you so far away from everyone?"

"That's the idea," said one of the insects.

"They're hoping we starve before they do," said another.

"Why?" said the fox. "What's wrong?"

They started speaking all at once.

"When I was young, everyone kept telling me it was beautiful union!"

"When the portals were open, I could go to other worlds," said someone else simultaneously. "It kept me distracted, but now that the portals are closed and everyone is isolating themselves, there's nothing to do!"

"It hurts!" cried one of the insects.

"It's not like they said! There's no harmony! I hate having another mind inside of mine! I hate my Guest! I hate being a Host! Why can't I live without this thing on me?"

"Why can't I live without her?" said one of the insects. "Why can't we ever agree on what to do? We keep arguing! We keep fighting each other! I want to escape, but I can't move! I can't move!"

The one Kylac had met outside the settlement spoke. "We never got along. My Guest... doesn't like me. I hate it. I wish we could get away from each other, and we used to handle the feeling by going to other worlds. It helped. Now we can't..."

The insect on his head spoke at the same time. "I can't escape! His thoughts are mine and my thoughts are his! We're not one! We're not one!"

"She makes me do things I don't want to!" said another insect. "I hate the food!"

It reminded Kylac of when he was an apprentice and accidentally spun a portal underwater. Everything had gushed out, flooding a whole area of Rel, and he forgot how to stop it. Here, emotion kept rushing out of these people, and Kylac took in all their voices. After several minutes, he realized they were repeating the same things over and over. Finally Kylac rose to his hind legs. "Everyone, calm down! Quiet!"

One by one, the Uiv became silent. They stood low, as if they wanted to burrow and hide from the insects attached to them. Some of the insects were flexing their legs and leaning to one side, as if they wanted to crawl away and hide.

"Have you spoken to anyone else about this?"

Everyone answered him at once.

"I did!

"They told me nothing was wrong!"

"They told me disharmony was impossible!"

"Once we are joined in childhood, everything will fix itself!"

"They are wrong! They are wrong! Something is wrong!"

"They won't listen to me!"

"It shouldn't be this way!"

"We don't speak together!"

"We don't think together!"

"Everything is separate!"

"It interrupts my thoughts!"

"—doesn't add to me! It's just riding on me!"

"I can't influence anything she does!"

And on and on, the felines acting more and more wounded, as if speaking this aloud physically hurt them. Finally Kylac raised his voice again. "Quiet! Quiet! Enough!"

They fell silent, more wounded than ever.

"Did any of you go to the Selts?"

"Many times," said the male Kylac had met outside. "They could not find anything wrong, and of course if the Selts can't find anything wrong, nothing must be wrong."

"It seems that way, doesn't it?" Kylac said, waving his tail. "And your healers found nothing, too?"

"They kept telling us it didn't happen," said an insect hidden in the group.

"It couldn't happen," said one of the felines, "because we are one."

That sent the felines into a flurry of snarls and roars. Kylac backed away a single step. He waited a few breaths for them to quiet down.

"I'm sorry," said the fox. "I don't know what to say."

The male Kylac had met outside the depression stepped forward. "After the other fox left, we had a thought. Your old ways."

"That's—"

"Starving isn't working!" someone screamed. "They won't let us kill them!"

"Even when they can't control us, they changed something in our brains so we can't hurt them!" said someone else.

"You can do it," said another. "You can kill them for us."

Kylac couldn't help but notice the insects had nothing to say about this.

"You're away from your raptor," said the male. "It's perfect."

Kylac stepped back. "Someone has to be able to help you. I'll help you find them."

He smelled someone behind him and leaped to the side just as one of the felines pounced on the spot where he stood. He turned around. He was surrounded.

5

The first kitten slipped out of the mother, and someone planted a tiny insect onto its head. The bug felt its way around, shifted, made clicking sounds. Then it drove its feet into the infant's skull. The infant screamed, mewled for a moment, then fell silent, insect and hairless lion unified. During this, Ogu and her insect narrated.

"They-used-to-take-over-our-minds-com-plete-ly. Cen-tur-ies-ago-we-fought-them. We-con-sid-ered them-pre-da-tors-be-cause-that-was-in-ess-ence-what-they-were. They-preyed-on-us-used-us-to-keep-them-selves-alive."

Another kitten emerged from the birth canal. An insect crawled onto its skull and implanted itself, the sound of bone breaking clear and loud. Stephen leaned on Deka as they watched, holding onto his arm as if his life depended on it.

"The-mam-mals-tried-to-ex-ter-min-ate-the-in-sects-by-des-troy-ing-much-of-the-land. They-di-ver-ted-ri-vers-to-de-stroy-moun-tain-sides-and-make-the-land-as-des-o-late as-poss-i-ble. It-did-not-work."

Two more kittens slipped out. A dozen insects crawled over her. Two found her skull and fought each other for position. One emerged victorious, shoved its feet through the skull, the infant screamed, and then fell asleep.

"And-then-one-of-the-in-sects-tried-some-thing-diff-er-ent. It-did-not-ex-tin-guish-its-host. It-lived-with-its-host-in-stead. That's-when-the-in-sects-real-ized-what-harm-they-were-do-ing."

Another kitten. Another bug. Another scream.

"When-they-stopped-for-cing-it-our-minds-be-came-one-and-we-learned-from-each-other. We-learned-very-well. It-be-came-no-long-er-a-matt-er-of-par-a-site-prey-ing-on-the-mam-mal. We-be-came-one."

Several cats came by and chewed the umbilical cords. Others ate the placentas.

"The-in-sects-be-gan-kill-ing-an-y-of-their-kind-that-took-ov-er-their-host-com-plete-ly. Now-it-is-in-stinct-for-both-spe-cies-to-grow-up-to-get-her. As-one."

Ogu held Stephen's shoulder, turned him around, and faced him the other way. So many people were crammed in here, and there were a hundred more layers just like this, housing their entire culture.

"Since-the-war-this-is-the-only-hab-i-tat-left-for-us-to-live," said Ogu and her insect. "We-use-the-por-tals-to-link-to-the-an-i-mals-we-hunt-so-we-don't-have-to-fo-llow-their-mi-gra-tion-patt-erns-any-more. We-al-most-starved-af-ter-the-dis-as-ter."

Stephen was shivering, and his skin felt cold. Deka kept an arm around him, but it did not seem to help. The human pulled Deka backwards a few steps and leaned close.

"Deka, can we get out of here?" he whispered.

"Why?"

"This place is freaking me out!"

"Why does it scare you?"

Stephen tried to speak, but nothing came out.

"You're not the only one who finds this terrifying at first," Deka said, loud enough for Ogu to hear. "Many learn to resent parasites, as they usually contribute to the downfall of a species."

Stephen was visibly worried that Ogu would hear him. He kept his voice low when he spoke.

"This is horrible! Did you hear those babies scream? Why did we come here?"

Stephen's breathing was erratic. On his breath, Deka smelled definite blood loss to his head, which meant he was not thinking.

"We need to go outside," Deka said to Ogu.

"Fol-low-me."

She began moving through the tight, thick crowd of hairless felines and insects. Deka led Stephen in her wake. Stephen did not let go of Deka's arm. He did not let the felines touch him, and Deka noticed he avoided so much as glancing at any of the insects.

A few minutes later, they reached the main entrance. It was always open, so they walked through, and Stephen finally let go of Deka and caught his breath.

"Feel-bet-ter?"

Stephen glanced at her, then stared at the ground again. Deka knew the expression on his face.

"You can say what's on your mind, Stephen."

"Thought you wanted me to be a good ambassador for my species?"

"Being a good ambassador does not mean hiding what you think."

"Really? That's different." The human straightened up, glancing by the door to make sure his backpack was still there.

"You-breathe-much-eas-ier-out-here," said both voices of Ogu. "And-you-smell-af-raid. Why?"

Stephen hesitated, then spoke. "I don't understand. How is this good for the cats? The bugs latch onto them, feed off them, do nothing, and you just accept this. They're dead weight! The cats do all the work, and the bugs just live."

"You come from a planet where survival by one's own means was rewarded," said Deka. "Even in groups. But what if survival by a different method had been rewarded? What would you do if you found out a parasite on your world was self-aware?"

"I'm sorry, but I can't see a dog learning to love his fleas. We just don't make friends with creatures that suck us dry."

Deka approached him, lowering his head to his eye level. "When you can leave your planet with the attitude that everything you know is wrong, then you are ready to explore the universe."

"In-the-be-fore-times-it-was-as-you-fear. The-in-sects-used-to-take-con-trol-of-their-fe-line-bo-dies. They-stamped-out-all-per-son-al-ity-traits. But-things-changed-when-we-real-ized-we-were-both-in-tell-igent. We-learned-to-live-to-get-her. As-one."

"You keep saying that! As one, as one! Don't you see it? The bugs are still in control of you! They're forcing you to keep them alive, and they give nothing in return!"

Deka was about to interrupt, but Ogu bumped him on the side. Stephen kept talking.

"This isn't an equal relationship! The bugs can't survive on their own, so they latch onto you guys! How do you know they're not making it so you can't fight back? How do you know they're not in control right now? They're already making you live like insects! This is just wrong—it's fucking wrong!"

Stephen was out of breath. He leaned on his knees, and Deka stood over him, placed a hand to his back.

"Part of learning about life off your planet is accepting life doesn't always work the same way. What seems wrong to you works very well for someone else, and you must be open-minded enough to understand that."

"But it's not so good for the cats," Stephen said, his voice a raspy wheeze.

"The insects are intelligent, living creatures. It's not their fault this is how their bodies work. Look at you, human."

Deka poked a claw into Stephen's back. Stephen winced and hopped forward a step. Deka held up one finger.

"Feel that? All I had to do was tap you and my claw went straight through your skin. Many species would remark you have no right to be alive because you are defenseless. You would resent that, but it's not your fault your body is like this, and yet you think you have the right to be alive. The Guests are no different."

Stephen turned and looked at Ogu. Deka noticed he was looking at the bug.

"Is-the-pan-ic-out-of-your-sys-tem?   Is-it-safe-to-take-you-in-side-a-gain?"

Stephen turned to Deka. Deka didn't know why Stephen still did it, since Deka had no facial expressions to read. He barely understood Deka's laughter.

Stephen nodded and walked toward the Uiv. Deka scanned across the land. Kylac should have been back by now. He had a suspicion something might be wrong, but he couldn't leave Stephen alone, not yet. He followed the human back inside.

6

"You need help!" Kylac shouted. "Go back to the colony and tell them what's happening!"

Claws raked Kylac's thigh. The fox yelped and stumbled into the waiting claws of another feline. She grabbed Kylac with one paw, raked her claws into his back and ripped away. Kylac screamed as he staggered backwards again.

He stood still in the middle of the circle of starving, hairless cats. All of them were snarling at him, growling, swiping their claws at him, not a thought in their minds.

Kylac stood on his hind legs in attack stance. He knew how an animal would behave, but these people were not normal, and he wasn't sure what they would do. Among many species, Kylac could defuse a situation like this with sex, but the cats wanted blood.

Some of them stumbled around, swinging their heads and paws. The Guests were trying to take control and force them to move away. Others clamped the jaws of their Hosts closed, and the felines struggled to open their mouths.

Kylac ran for the closest disabled feline, shoved him out of the way, and plowed through. The feline behind him had not been disabled and reached out to slash the fox. Kylac jumped above her, landed on someone's back, leaped off and landed on all fours, about to dash for the water—

Kylac felt his tail stretch out, and his body snapped back. Paws walked up his spine, and enormous weight lay on top of him.

"I'm sorry! I'm trying to stop her!" an insect was saying.

"Quiet!" the feline snapped. "This is why I'm here! I can't rip you off my head myself! Help us, Kylac!"

Kylac's chest was so tight he couldn't breathe, but he found a voice. "Get help!"

"Stop!" shouted the bug. "Stop! Stop!"

"Be quiet!"

She extended her claws and gouged them into Kylac's back, pulling down to his hips. Kylac couldn't even scream.

"Help us!" someone shouted.

"I want my own thoughts!"

"I want control of my body!"

She raked her claws down Kylac's back again. Kylac was out of breath and couldn't inhale.

"Kill them! Kill them!"

The cat on top of Kylac wrapped her jaws around the fox's neck and squeezed. She could pick him up and snap it if she wanted, and only this knowledge kept Kylac calm. After a few moments, she let Kylac up. Kylac stood, choking and gasping.

"Kill them!"

"Silence them!"

A few of the insects voiced their opinion by forcing mouths to close and compelling some of the Hosts to back away from the group. The felines were shaking their heads, trying to wrestle control back.

Kylac turned in a circle, searching for a way out. Someone stepped close to him. Kylac kicked him in the snout and pushed him back, but another one leaped close to him. Kylac spun around, raking his claws across his snout, but he bowled Kylac over and knocked him onto his back. The fox made one last effort to avoid this. He raised his legs, exposing himself, but the feline clamped his teeth on his snout, picked Kylac up and threw him into the mob.

Felines descended on him, even the ones who couldn't open their mouths. Kylac raised his arms, scratching and biting as many as he could. He snapped his jaws at many more and batted a few of them away, but others always moved in, and claws dug into him. Kylac smelled blood, his and theirs. He smelled anger and frustration, but this was not impulsive. This was rage that never had a release until he showed up. Rage that had never been expressed before. Rage they had denied for their entire lives.

Kylac was pinned under the oppressive scent of pent-up frustration and hate. He thrashed and he flailed, scratching anything he could, biting anything his mouth would reach. Some of them angled their heads so his claws would hit their Guests. He scratched one of them. The bug writhed in agony, and so did the feline.

He hoped that would be enough to dissuade them. Instead it emboldened them. They wanted it to hurt. They wanted the insects to suffer. Kylac thrashed harder, trying to break free as they bit and scratched him.

Then five of them lay on him, one on each limb and the other on his torso. Kylac lay splayed out, still trying to figure out how he could convince them to have sex instead. He imagined how it would feel to encounter a ferocious pack of hungry, snarling cats, and Kylac was able to entice them to fuck him one after the other for an entire day instead until everyone fell asleep, satisfied, and then they would keep him around to dissolve tension in the group, using him again and again every day—but then the claws and teeth hit him, and he couldn't escape into fantasy.

One of them seemed to have been hurt the worst. He stood over Kylac, letting his blood drip over Kylac's muzzle. Kylac turned his head to avoid it, but the feline on his chest held his head in place between her enormous paws.

Everyone stopped moving for many breaths. Kylac tried to ignore the blood dripping on him. He meditated on portal calculations to Friend's next destination, just as he had done when he was a young apprentice and had to meditate for days in total isolation. He concentrated on the true nature of the universe. He hoped the cats wouldn't notice, and he swore he would never leave Deka's side again for any reason—never again, never, never, never.

7

The insect crawled up Stephen's arm. It was a fast mover, and its legs felt sharp, as if about to poke through his skin. Though it had a mouth, it did not use it for eating, but rather tasting for a host. The young insect was doing just that to Stephen's skin right now. The human had frozen up, too scared to shiver. Deka kept a hand on his back to keep him calm.

"It's harmless, Stephen. It won't attach to anything but the infant cats on this world."

It was moving up, up, up Stephen's arm. It reached his head, extended a tube from its mouth and felt Stephen's hair. It reached up, crawled up Stephen's head. Now Stephen shivered.

"Calm yourself."

Ogu sat in front of Stephen. She held her head in a position expressing curiosity: tilted forward, looking at him from the top of her peripheral vision. It seemed to make Stephen nervous. Even though Deka had already explained what the gesture meant, the human still reacted to it as though he were being hunted.

The insect had now perched on top of Stephen's head. It was cute to watch how it reacted to the hair. The Guest had never felt hair before, and it probably never would feel it again.

It held still. So did Stephen, as if any move would make it bite. Deka wondered what remnant of survival instinct prompted this reaction.

The insect climbed down from Stephen's crown and tasted around his other shoulder. Deka reached over, picked it up, and held it by the belly in one hand. All six legs flailed around, trying to find the ground again.

James L. Steele

"Without a Host," Deka said, "it will only live for about a week. It has no way to take in nutrients. No digestive system. No mouth connected to a stomach."

He held one of the legs up to Stephen.

"Its legs are actually its mouths. It can only take in dissolved nutrients already in the bloodstream. Extensions of its nervous system also run down its legs, which joins it to the neural pathways of its host."

Stephen stared at it. Then he turned to Ogu, facing the Guest on her skull.

"What would you do if you had no digestive system, Stephen?" Deka let go of the bug and allowed it to crawl up his own arm. It perched on his shoulder and felt around. "If you couldn't survive by your own means—if you were physically incapable of living on your own, what would you do? How would you have to live?"

"I get it," said the human. "But it still kinda freaks me out."

"Man-y-spe-cies-have-that-re-act-ion," said Ogu and her Guest. "You-are-an-is-o-lated-spe-cies-so-I-un-der-stand-why-you-would-think-it-is-a-bad-thing. You-can-not-feel-what-we-feel. The-Guests-gave-up-a-lot-so-we-could-live-to-get-her. They-gave-up-their-to-tal-con-trol-o-ver-us-so-we-could-live. They-learned-how-to-live-with-us. We-learned-to-trust-them."

"But..." Stephen stammered. "They're still feeding off you."

"They-have-no-choice. They-will-die-if-we-do-not-help-them-live."

"I'm sorry, I still don't see how that would be bad."

Deka suppressed a grunt, relieved that Stephen had finally said it. "Are you hungry, Stephen?"

"Yeah. Thirsty, too. This place is hot."

"So am I. We'll go to the hunting grounds."

"Can I eat it uncooked?"

"Have I ever told you to eat anything that made you sick?"

Stephen nodded. "Right, let's go."

The Guest climbed down Deka's back. He lowered his tail, and the insect crawled off him. Deka didn't worry about what would happen to the little one; someone would take care of it.

They squeezed through the thick convention of felines toward the portals against one wall. Normally the outer wall of each layer would be full of spheres. Now only a few led to the hunting grounds. Ogu took the lead and walked through one of the portals on the left. Deka followed, leading Stephen through.

They emerged on a narrow valley between two jagged mountain ranges. A herd of horse-like creatures grazed in the distance, and several Uiv walked the perimeter, keeping the herd together while stalking the prey.

Ogu walked them in the direction of the herd. The grass was short, but abundant. Stephen was looking at it.

"Funny how there's some kind of grass on almost every planet," Stephen observed.

"The-sim-ple-est-forms-of-life-are-of-ten-the-most-suc-cess-ful. This-is-true-on-ma-ny-worlds."

"This stuff is softer, though. Why is there grass here but not where you live?"

"Be-cause-of-the-war.   Be-fore-the-Guests-knew-we-were-in-tell-i-gent-the-Hosts-tried-to-kill-the-Guests. They-ru-ined-much-of-the-land-where-the-Guests-lived.   The-land-still-has-not-re-cov-ered.   The-an-im-als-we-hunt-still-mi-grate-be-tween-the-des-ol-ate-parts."

"You mean these pastel ponies?"

"The Uiv simply call them the prey."

He laughed. "They look like they rolled in sidewalk chalk!"

"I-do-not-un-der-stand-what-you-mean."

"They just look funny. Similar animals live where I come from, but they're all earth-toned."

"They can run up to thirty miles per hour," said Deka. "Good luck catching one."

"Definitely."

"I mean it, Stephen. Go catch one."

Stephen glared at Deka as they walked. "You're serious."

"I have caught all the live meat for you and Kylac over the last three months. Kylac has been finding all the fruits and vegetables that are safe for you to eat. You have given nothing in return. Let's find out how you survive on your own. Ogu..." Deka switched to her native language and told her to come with him.

Deka took off running, and Ogu ran after him at full speed. She could match Deka in a sprint, but not long distance, and quickly she fell behind. Deka did not wait for her, or for Stephen, even as the human shouted at him.

8

Blood saturated the fur on Kylac's face, and it was starting to smell good. Kylac had reached a point where he didn't want to meditate on sexual fantasy anymore. He wanted more of that smell. Blood was good. If something was bleeding, that meant it was wounded, and anything wounded couldn't harm him. The chain of logic required no thought.

Kylac tried to think of Deka, tried to keep him in mind. He didn't want to disappoint him again, but the blood tasted so good...

"Everyone... You don't know what you're doing. I can't control myself. I'll kill all of you."

Nobody responded. They thought they could control him, use him, and he hoped they were right. Kylac couldn't hold back anymore. He surrendered. The blood dripping over his muzzle smelled too good. He opened his mouth and let it fall on his tongue.

The felines became more attentive. Kylac felt teeth around his wrists, teeth around his forearms and biceps. They lifted him off the ground, and the bleeding feline moved away. Kylac was now out from under the blood, and the fresh air was a relief. He tried one last attempt to clear his mind, but it was too late. He loved it. He wanted more...

...and there were so many scents around him. Scents made him angry, and he snarled at them.

The jaws propped him upright on his knees. Kylac leaned forward and snapped his teeth at the felines holding his left arm. They responded by clamping down. Kylac took the hint and faced forward. They loosened their grip but held him still. Blood ran off both of his arms and dripped from his legs, too.

Kylac loathed their scents. He snarled and snapped his jaws at the air between him and the surrounding felines. Kylac forgot how to speak. The world was made up of scents he had to extinguish, and he wouldn't be calm until his nose was clear.

One of the felines approached, holding her head up to him. Kylac snarled and barked louder. She came within striking distance, and Kylac scented the bug. It was making noises, and Kylac hated both the scent and the noise it made.

He opened his jaw, tore into its back, and ripped it open. The feline screamed all the way down to the dirt. Kylac tried to finish the kill, but the others held him firmly in place. He struggled in their grip and growled at them. They clamped their mouths tighter, and Kylac held still.

The female slowly, painfully stood up, and then offered her head again. Kylac opened his mouth, grabbed the entire insect, and tore it off. No blood, only insect juice, which disappointed the fox.

The feline went down again, nothing left of the bug on her skull except six twitching legs sticking out of her head. The feline rolled away, curled into a ball, and screamed. Kylac liked hearing her in pain; it meant she would be no threat.

Another feline stepped up to Kylac, offering the insect on his head. Kylac refused at first; he wanted blood, and these bugs had none. The felines holding his body down thought otherwise—they bit down harder on his arms and legs, making Kylac snarl again. He lunged, clamped his jaws on the bug, and tore it off. The cat wailed and crumpled on the dusty ground.

Kylac now enjoyed taking the bugs off. It hurt the felines, and they were no threat to him without their Guests. He waited patiently, growling, drooling in anticipation.

Another feline offered his head. Kylac tore the Guest free and swallowed it. It was starting to taste the way blood should. The male stumbled away, reaching up and trying to rub the legs off of his skull, reeling from the pain of having a part of himself ripped away and eaten.

Feline after feline. Insect after insect. Kylac ate a few of them, but most he just chewed up and tossed over his shoulder. He snarled at them to hurry up, and the Hosts couldn't line up fast enough.

Finally everyone in sight was free of their insects. Some still writhed on the ground, moaning in pain. Some seemed unable to catch their breath. Others couldn't seem to control their bodies anymore.

But quite a few were up and about, functioning perfectly now. This angered Kylac—He hadn't hurt them enough and now they were a threat to him again! He snapped his jaws at them and snarled.

Kylac felt teeth loosening from his leg, and then a new set of jaws clamped him. One of the felines holding his right arm released him, and a new jaw closed around his arm in the same place. The felines traded off, and now Kylac liberated the cats who had been holding him.

Some of the felines lay on the ground. They smelled dead, which made Kylac happy, but most of them still stood, not bleeding, which enraged him.

Kylac vaguely recognized the feline standing in front of him as the one he had met outside the clearing. He was covered in bug guts, and the blood vessels on his skin were so raised and red he appeared to have no skin at all.

"Thank you. You did the one thing we could never do. You may calm down now."

Kylac snapped his jaws at him and snarled. His scent was loud in Kylac's nose, and he had to make it quiet. He growled and screamed at him, tried to move his arms to slash his eyes, but the jaws held him still.

"Roll him back."

They pulled Kylac backwards. The fox resisted, but the tight jaws forced him down. The jaws released his legs, and Kylac tried to kick, but quickly two more felines clamped their mouths around his legs and spread them apart. Kylac snarled at them. Another held her jaws around his neck and pushed his head down. Kylac could only growl quietly now.

"Calm down," said the male. Kylac felt him climbing on top of him. "Just calm down."

Kylac howled.

9

Stephen watched the herd move about. They observed him but did not react to him at all. Instead they moved as the Hosts moved them. They coordinated, brought certain groups together, pulled others apart, separated them into smaller and smaller herds. Stephen watched a group of three naked lions take down one of the bright pink horse-things. It had a blue tail and white flecks running up its neck. Stephen couldn't help but smirk at just how absurd this looked, and yet it was life and death.

The three felines converged on the horse and tore into it. The herd scattered. In the panic, the fifty other felines brought down horses of their own with barely a struggle. Some gorged themselves right there, others carried the prey back to the portals.

Stephen watched them, shaking his head. He held his multi-tool at his side, knife out. Stephen had gone back to fetch it, getting crushed by the hive of insects and giant cats. Now his backpack lay by his feet, and it felt good to wear clothes again. He didn't think very long about this, only that if he had to hunt, he should be comfortable.

His clothes were dirty, and so was he, but Stephen didn't care. He hadn't realized it until now, but he was tired of being told he was doing everything wrong. So what if he didn't have fur to keep him warm? He had clothes. So what if he didn't have claws? He did have a knife, and that was more than enough for him. He had to accept others for what they were, so why couldn't they accept him for what he was? It wasn't his fault he had to survive like this, but it worked.

He held no hope of running up to a creature and catching it. Instead, he approached the herd cautiously. He wished he had some kind of camouflage, but it wouldn't do much good out here in the open grassland.

Several horse-things in the herd stared at him as he approached, knife ready. Stephen hoped they wouldn't recognize him as a threat. He was still about twenty yards away when the herd galloped away from him. Stephen lowered the knife and sighed. A few of the lion-creatures snorted at him and chased the herd. Stephen followed, thinking it felt good to run in shoes again.

The herd stopped several hundred yards away, and Stephen tried again, this time moving slower. He got to within thirty yards, and then the herd of horse-things fled. A few felines approached, made dual-voice sounds at him from both their mouths, and chased the horses. Stephen got the feeling they didn't like him ruining their kills. Stephen wished he could have brought a gun.

He chased several herds, never getting within ten yards, and by the time the sun set, he was leaning on his knees, panting. Stephen looked over at one group of felines devouring a kill. He didn't want to do this, but he didn't seem to have a choice.

He walked to the group. They regarded him silently. Stephen took a deep breath and sat down among them. He held the knife up, slipped it in the prey, glancing around at everyone again. They were still staring at him.

Stephen carved a small piece and held it up. It smelled like raw beef, which wasn't bad, but he still wished he could build a fire. He took a bite. It tasted like raw hamburger, but it was still warm, so it reminded him of rare steak. He hoped he wouldn't get sick.

One of the cats clawed a hunk of meat free and held it out to Stephen. Stephen almost stabbed it with the knife, but thought better of it. He took it with his bare hand and bit into it.

Stephen nodded in thanks. The felines offered him a few more bites as they continued making sounds at him, half coming from their mouths and the other half coming from the insect's.

Stephen smelled reptile. Deka spoke from behind him.

"They saw you try to hunt. They know you can't, and they understand that. Believe it or not, they respect it."

Stephen stood and stretched his neck to try to be at Deka's eye level. "Listen, Deka, that's starting to get insulting. So what if I don't have fur? So what if I don't have claws? So what if I need a gun to hunt? I can't help that! My people are a product of tools, but so what? It's how we have to survive! Why do you and your fox keep telling me I should be ashamed of that?"

Deka curled his neck as he lowered his head and held Stephen's stare. "Because you still believe you are better than the Guests because you digest your own food. Take away your tools, and you are no different."

The human did not avert his gaze. "So we need things. Who doesn't? They built that tower! Those lizards on Loam build rock piles! I don't see how it's a bad thing."

"Everyone here can hunt. Every Loam reptile can build those rock structures. Every Ven on Hithe can control the rock growth. Each feline on Selta is a healer. It's what their kind does, and they're using those abilities to help each other rise above their animal nature. You can't make a gun. You can't make a trap, or a car. Even with the tools you did bring, you can't fend for yourself. You have placed your value in something you cannot make."

Stephen fumed. "We're a community species! I learned that in school for Christsake! We rely on each other to do different things! That's our fucking nature!"

"Your kind is using tools to entrench people in isolation, mentally and physically. You have been kept helpless and dependent on things you cannot make on purpose, Stephen, and that's why it is not something to be proud of. Tools reinforce your primitive nature instead of helping you rise above it, and you have relied on them for so long you can't survive without them, which makes you a parasite. Do not despise the Guests. Remember why they gave you the food."

Stephen looked back at the felines.

Deka continued. "You can't tell, but they like you because you can't hunt. That's what happened when the Hosts discovered the Guests were intelligent. It filled them with compassion. The gesture went both ways when the Guests discovered their Hosts were intelligent. They stopped being enemies, and sympathy replaced fear."

Stephen shook his head, turned and met Deka's eyes again. "That's not sympathy. That's pity. I'm a lot of bad things, but I'm not pitiful."

Deka held his claws together. "I knew this planet would test you. You value self-sufficiency because it is the one thing you are not, Stephen. Incredible to think you can believe you are one thing when you are in fact something else. This is what your species lacks. Without a companion, you have no idea who are you are. You will not be able to know the universe the way it is until you can understand yourselves from someone else's point of view. If humanity is ever to join the contacted universe, it will have to become used to seeing things it considers disturbing."

Stephen couldn't hold the raptor's gaze anymore, and he turned his head to the felines. Their Guests stared back at him. He shuddered.

"Come with me," Deka said, raising his head to full height. "I have your backpack. Let's find Kylac."

10

Deka, Stephen, and Ogu walked over the hill where Kylac had gone off to pursue Friend's scent trail. Deka and Ogu bent down and scented the rocks. Stephen stood still and watched, fully clothed and grateful to be wearing shoes while standing on the sharp rubble.

After much searching, Ogu and Deka found Kylac's scent and were finally able to follow him away from the tower. Stephen hopped from rock to rock and tried to keep up.

Deka and Ogu walked side by side. The raptor paused every few paces, picked up the scent, and kept going. He could just barely get a whiff of Friend's odor, but Kylac's was much stronger because it was more recent.

Ogu spoke to Deka in her native language. "Will-Stephen's-spe-cies-leave-their pla-net?"

"Probably not."

"So-he-may-be-the-on-ly-hu-man-I-will-e-ver-meet."

"Likely."

"For-an-i-so-lated-spe-cies-he-han-dles-him-self-ver-y-well-off-world."

"He has done well. I hope he learned something. I fear he will simply forget everything when he has to return."

Deka looked back to make sure Stephen was still behind him. He scented the air, caught Stephen's scent and noted the human was much calmer away from the Uiv.

"We gave him a glimpse of what real aliens are," he continued. "I hope it makes him happier, but I think it will only make him miserable."

They followed Kylac's scent over the hills of rubble and between stagnant ponds. Ogu walked wide around the pools, not wanting to draw the attention of any biting insects.

Ogu stopped and focused on one particular area. So did Deka. He smelled a Host. Kylac had met someone out here.

"Ogu... People live outside the tower?"

"No. No-they-do-not! I-do-not-know-this-scent-and-I-do-not-know-who-he-would-have-met-out-here.   But-he-smells-dis-turbed."

Deka walked ahead, nose to the rocks. "Kylac went with him."

Deka took off running for a depression in the ground. Ogu followed on his heels. Stephen was still far behind, carefully hopping over the rocks.

Deka and Ogu reached the ledge. The scent path descended into an area where the rocks had been cleared all five paces down to the dry soil. The bodies of twenty-seven Hosts lay in the depression below, most with insect legs sticking out of their skulls.

A stagnant lake filled the far end of the depression, and a fox was kneeling in front of it. He was covered in blood. Stephen had just caught up to them.

"Deka, is that—?"

The theropod ran down the slope, touched down on the bare dirt, and dashed toward the fox.

The fox heard him. Deka saw what his nose had already told him. His fur was so bloody there was no white left on his body. The fox raised his lips and snarled at Deka as he approached. He rose to his hind legs, held his claws up, teeth bared.

Ogu had climbed down the jagged rocks to the soil. She stopped at what she felt to be a safe distance, about twenty paces away.

When Deka was within ten paces of him, Kylac charged. Deka let the fox ram into him, caught the canine in his arms and clutched the fox tight against his chest and belly. They slammed to the ground together and rolled until Deka lay on top of Kylac. The fox struggled and snarled and screamed at Deka. The raptor smelled the blood of many Hosts on his body, and his breath smelled like Guest.

"Kylac," Deka said into one ear.

The fox was still thrashing and snarling, but Deka's entire body held him down. Deka looked up, noticed Stephen perched on the ledge.

Kylac kicked a claw into Deka's slit. Deka screamed. Kylac kicked again. Deka's grip loosened, and Kylac wiggled free, leaping on top of Deka and sinking his teeth into his shoulder. Deka rolled back and kicked Kylac away. The fox sailed through the air and landed on his back, skipping across the ground and rolling several times before stopping and rising to all fours.

Ogu was backing away. Kylac snarled at her, flashed his teeth, and charged. Ogu turned her head, shielding her Guest. Deka dashed and rammed into Kylac just a few paces away from Ogu, throwing him to the ground. He tried to lie down on the fox again, but Kylac rolled and leaped away as the raptor scrambled to grab an ankle or an arm or a handful of fur.

Kylac saw Stephen at the top of the ledge, breathlessly staring down at the fox. He backed away. Kylac screamed at him. Stephen spun around and bolted, and Kylac scrambled up the rocks. Deka ran to catch up, Ogu right behind him. Kylac was already to the lip by the time Deka reached the bottom. Deka jumped, cleared the height of the depression and landed on the rocks, already dashing after Kylac.

Fifteen paces away from the ledge, the human and the fox squared off. Stephen was holding his knife ready, shouting obscenities at the canine. Kylac was snarling, growling, slashing the air between them. Deka closed the distance, hoping his feet wouldn't slip between a crack in the rocks.

Stephen stumbled, glancing down frequently. Kylac charged. Stephen slashed with the knife, but Kylac collided with him, knocking it out of his hands, sending him down to the jagged ground. Kylac pinned Stephen underneath him as the canine sank his teeth into the human's shoulder.

Deka leaped long. He landed on the rock right over Kylac's head and closed his jaws on Kylac's neck. He lifted Kylac off Stephen and threw him on his back. As Kylac screamed, Deka lay on top of him, wrapping the fox up in his arms and legs.

Kylac squirmed and struggled, snarling and howling and making noises that had not been part of the language since the two species united. Deka left him no room to move, forcing him to breathe Relian scent. His muzzle was right next to Kylac's, blocking its movement as well.

"Kylac... Tell me what happened. Find words again. This is not you. Calm down."

Kylac still howled. His voice would be audible all the way to the tower.

"Come back. Come back. Come back, Kylac."

Out the corner of one eye, Deka saw Ogu standing over Stephen, licking his shoulder wound. Stephen was trying not to scream, and he was shivering, wanting to run away from all of this, but pain paralyzed him.

"Kylac, this is not you."

Kylac growled and tested Deka's weight again. Deka was worried now. Usually just seeing him was enough to bring him back, but this time he was too far gone. Deka did not want to do this. He closed his eyes, took a quick breath, craned his neck backwards and brought his mouth down on Kylac's neck. He stood and brought his foot down on the canine, shoved his killing claw into his leg just a little, and slashed. Kylac screamed again. Deka leaned down, raked his hand-claws across Kylac's chest, slowly.

Kylac howled and tried to wiggle away. Deka stood on top of him, shoved both killing claws into him, not very far, but enough to pin Kylac down. Finally the fox stopped snarling. Deka waited. He surrendered. Deka lay on top of him again, wrapped him up tight and repeated himself over and over.

"Kylac. Come back. Come back. Come back."

Deka waited. Kylac didn't scream or struggle anymore. His scent began to relax at last.

"Remember. Remember."

Kylac's scent had been hot in Deka's nose, but now it cooled. Deka kept a good grip on him for a few more breaths.

"I'm sorry..." came a ragged, exhausted voice from beneath him.

Deka released Kylac and sat on the rocks beside him. The fox sat up and faced Stephen and Ogu. Stephen was sitting nearby, shirt torn off and cast aside on a rock. Ogu was nursing a nasty mouth imprint over the human's shoulder. His back and side were badly bruised. Kylac gasped for breath, reached out to Stephen and lay a paw on his chest just below the wound.

"Stephen, I'm—"

Stephen smacked his hand away and climbed to his feet. "Get the fuck away from me!"

"Please, I can—"

"Fuck off!"

He made sure his backpack was still on and stomped away. Ogu was about to follow, but Kylac held her back.

"Ogu..." said the fox, still in English.

She stopped and regarded Kylac. Deka leaned over him and began licking Kylac's wounds.

"They were living here," said the fox. "Hosts in disharmony with their Guests."

"Dis-har-mon-y?"

"They did this to me. They wanted me to kill their Guests for them so they could be free. They were Hosts... who didn't join properly. Without the portals offworld, they had no distractions from it. It drove them insane. They held me down. They forced me to kill their Guests. But they forgot... the Guest integrates its nervous system with the Host, including heartbeat and breathing. Most of them died on their own. A few lived, and they thought they could bring me back. I broke free before they could try. I killed them. I'm so sorry. Ogu, I'm..."

She looked out over the depression. Her head hung, and she began to shiver. She coughed. Her Guest also began coughing as it shared the grief.

Kylac turned to Deka and licked the blood off him. Deka didn't allow him to taste blood for too long and nudged his muzzle away from his open wounds. Kylac sat still and let Deka clean him, shivering.

"Help them," Kylac said. "There are more. They're ashamed. People here pretend disharmony between Host and Guest can't happen, but it can."

Ogu still stared. "But-it-can-not-hap-pen. Dis-har-mon-y-is-im-po-ss-i-ble. We-are-one."

"Sometimes things go wrong. I'm so sorry. Please. There may be others. Help them."

Ogu tilted her head. "I will find out who's missing and who knew them. Maybe someone knows who they were and what happened."

Deka stood and helped Kylac to his feet. His fur was still matted in blood, dripping from many places across his chest and back, but the old ways had calmed. Deka still bled from multiple slashes and bites.

"I'll get Stephen," Deka said, turning.

"No!" Kylac reached out, grabbed Deka by the neck. "Don't leave me!"

"You're fine now. Stephen is not. He's had a very bad time here."

"Deka, please!"

Deka stood still as Kylac held him. Deka held him back.

"I'm sorry. I'm so sorry." He whined as he shivered. "They forced me. I couldn't stop. It felt so good. I never want to feel that again please don't leave me."

They stood together for some time, Kylac mourning for what he had done. After many breaths, he finally calmed enough to speak again.

"Rive and Friend went to Kronia."

"I'll make the way. Stay with Ogu."

Kylac was ready now, and Deka let go. The raptor turned and saw Stephen some distance away, sitting on the rocks, huddled into himself. Deka trotted away from Kylac and Ogu, making sure his claws clicked against the rocks so Stephen wouldn't be startled. He stopped a few paces behind the human.

"Stephen."

"What!"

"We need to clean that wound."

"Fuck you and fuck this goddamn place!"

Deka chirped, a gesture of apology he knew the human would not recognize without explanation. "Kylac has animal impulses I've been helping him tame since we were children. I'm very sorry you were caught in the middle. The danger is over now. Kylac is calm again. We can explain more once you heal, and we don't have to send you home right away. There is one more place I want you to see. One more person you should meet. But it's your choice. If you want to go home now, then I will work on a way to Earth, as well."

Stephen sat in silence for a long time, gazing up at an alien sky, taking deep breaths, bleeding profusely from his shoulder. Finally he stood up and turned to Deka.

"One last planet."

Deka rubbed his claws together, walked up to the human, held his good shoulder, and guided him back over the rocks to where Kylac and Ogu waited. Stephen didn't even look at the fox. Kylac opened his mouth, but Stephen was not listening; the wounds were too fresh. Tail between his legs, Kylac stumbled over the rubble. The four of them walked back to the tower, leaving Stephen's torn shirt and his multi-tool somewhere between the cracks in the rocks.

Kronia

The portal opened into a large, empty cave. Deka, Kylac, and then Stephen jumped through. The human wore long underwear, coat, hat, thick gloves, and shoes. The wind blew in through the mouth, but no snow made it inside. The portal closed behind them.

"Damn, you weren't kidding," Stephen said. "This place is the north country. Been lugging winter clothes around for three months and finally you take me someplace cold."

"Deka doesn't like cold places. This is my kind of weather."

"Is the entire planet like this?"

"No," Deka answered. "Kronia has many different climates. Most worlds do. This is where Norh happens to live."

"So what is he, a ball of fluff?"

"Not even close," Deka said. "I'm curious how his scent affects you."

"And his appearance," said Kylac. "This planet seems perfectly in-tact. I'm opening a way to the hub. Maybe I can pick up Friend's scent."

"I'll go with you. Stephen, I want you to wait for Norh here. Kylac will keep a sphere open as long as he can so if you need to warm up, you can step through to the hub for a while, but if you see the portal start to shake, get to this side. I want you here when Norh arrives."

"All right. Jeez, it's cold. Aren't you cold, Deka?"

"Very."

Kylac's tail wagged wildly. "Deka has been trying to resist the cold for years. He wants to earn Norh's respect. You'll understand why when you meet him."

"How can you resist the cold?"

"Raptors learn to control their metabolism when we go offworld," Deka said. "I am not as good at it as some."

Stephen chuckled. "I thought you were good at everything. So what is Norh? What's his planet's story?"

"I'll let him show you," said Deka. "He'll be eager."

"Remember to keep talking to him. He'll learn your language. He won't hurt you, but he will be curious, so expect some probing."

"Probing?"

"You may have to strip."

"Here? It's gotta be fifteen below!"

"Very good," said Kylac. "On your temperature scale, it is fifteen blow zero. If you were out in the wind, it would be about thirty below."

"Just like home. Norh likes this weather?"

"The Krone aren't affected much by the weather."

Deka turned to Stephen. "I'm working on a way to Earth. Kylac will make the way to the next world if we find Friend's scent again."

"Okay."

The Relians sat on the edge of a large slab of smooth stone. Stephen sat down between them, looking from one to the other.

"So this is it? My last planet?"

"This time it will be," said Deka.

"Listen, I'm sorry how I acted back on Uiv. I was kind of a jerk, but there was something about that place. It brought out the worst in me."

"I do not believe that was your worst," Deka said.

"No, it's not," Stephen admitted. "I've done worse things. So I should probably ask some things that have been bugging me before I lose my chance. Like, why do you two have thumbs?"

Deka and Kylac raised their hands and examined them, wiggling their opposable digits as if only just now noticing them.

"I know why I have them," Stephen continued. "They're left over from when people lived in trees. Why do you two have them?"

Kylac wagged his tail. Deka rubbed his claws.

"Good question," Kylac said. "In my species' case, it had to do with being able to grip prey. I can walk on my hind legs because of the grass. Helpful to see over the grass and scent for anyone."

"And I have thumbs because we needed to hold on to foxes," Deka said. He wasn't laughing.

"All right. Makes sense. And Deka, you told me, take away my tools, and what am I? I've been wondering. Take away your portals, and what is everyone else? You told me lots of species almost died because the portal network collapsed. They built their society around them. That's something they can't produce. So. Tools. Portals. What's the difference?"

Deka considered the question for a moment. "Nobody expected something to happen to all the Archeons at once. It should have been impossible. The disaster killed so many people and ended many civilizations who relied on portals. Some did use them the way you use machines. There is an idea about that. The accepted theory is the contacted universe is in transition. Someday everyone will be able to do what Kylac and I can do, then it won't be a problem. Everyone can base their society on portals, and there won't be dependence on Archeons to create them."

"Maybe Earth's in a transition, too. Maybe technology will bring us to a point where we can make contact with others."

"It certainly won't. Machines replace human thought instead of enhancing it. It only makes your society more dependent on them, which further isolates you."

"How so?"

"That is one thing holding your kind back, Stephen," said Kylac. "Using an external device to replace your thinking renders your minds less capable. They have atrophied because they do not need to work. You think you are expanding your potential with technology, but you are actually shrinking it. Outsourcing intelligence only makes the individual less so, and thus the entire population. Mature civilizations know that each person relying on his own abilities is what releases a species' true potential."

"Remembering things is what leads to portal physics," Deka continued. "It cannot be recorded or automated by machine, so if you keep writing things down and letting machines do the math for you, you will never learn what your minds are capable of. You will never discover how the universe works."

"Portals make entire planets dependent on them, and it seems to bring everyone else up higher," Stephen said.

"Portal physics frees people from the burden of survival," Kylac said. "Humans have used machines to increase this burden."

"I get it now. Man, before computers, things took longer, but we still got them done. Then computers came, and they said everything will be easier and we'll have more free time. Instead, everyone just expects us to do more work. Some progress."

"Exactly," said Kylac. "If your planet discovered portal physics today, how do you think it would be used?"

Stephen laughed. "We'd open ways to every country and use them to station troops everywhere. We'd guard the ways with soldiers. We'd use them to speed up our commutes to work, and the airline companies would fence them off and make us buy tickets. Soon we'd all be required to be on call twenty-four-seven."

Kylac waved his tail as he spoke. "The machines you make are supposed to bring you together and make life easier, but humans are using them to relieve the burden of survival from a minority of people by shifting it to the majority. You would use portals in the same way."

"Yeah, Deka told me that already. I get the message."

"Any other questions?"

"Yeah. Why is everyone in the universe bisexual?"

Deka and Kylac both laughed.

"You still think you are not?" the raptor said.

"I'm not. I'm straight. Was married, about to have kids."

"That's not what you said in your bedroom." The canine wrapped an arm around Stephen's shoulders and held him close. "Or when you started kissing me on Celta and Neben, and all the other nights you spent with your dick in me, and let's talk about that night on Faii you fell asleep with my dick in you."

Stephen blushed as he looked at the ground. "You make me do strange shit, Magic Hands."

"There is a huge gulf between what you think you are and what you really are," Deka said, smiling with his hands. "From the entertainment we watched, I presume it extends to the entire species and not just you."

Stephen was silent for a moment, looking sideways at Kylac. He was touching Stephen's bad shoulder, and it ached. "What happened to you on Uiv?"

Kylac's ears folded back, and he lowered his arm, stared ahead at the cave entrance. "Are you ready to talk about that?"

"I may never get another chance. What the hell happened?"

Deka answered. "Before the two species of Relian became self-aware, the canines lived in the northern regions, where the climate was much colder. The theropods lived in the central regions, where it was warm. Raptors were driven by family ties. They survived the hardships of their environment by binding together in small groups. The canines lived in a much colder climate, and the limited resources forced them to become territorial. They learned to survive by keeping everyone else away from them. It made them aggressive beyond reason."

"The raptors tamed us," Kylac continued. "They dominated the canines until we submitted to them. They had to. We would have destroyed every living thing on the planet if they hadn't. The raptors controlled our breeding, choosing the least aggressive ones to bear young. It took generations, but eventually they had a population of canines whose territorial aggression could be converted into sexual desire."

"We call them the old ways," Deka said. "It's what a fox is without a raptor. It's why Relians are always in pairs. The raptors are the only thing keeping the foxes from living like that again."

Stephen took a breath. "Shit."

Kylac's tail brushed the back of Stephen's coat. "I am prone to reverting. When it happens, I feel scent anxiety. Any living scent in my nose that isn't bleeding makes me angry. Blood smells good because anything bleeding is wounded, and anything wounded can't hurt me. Just having a raptor around is usually enough to keep a fox's higher mind dominant, but even in these times it often requires force. Raptors took advantage of our instincts to tame us. You saw it on Uiv when Deka brought me back. When a fox feels defeated, he will surrender. It helped my higher mind reassert itself."

"So you have sex instead of attacking people?"

"Yes," said Kylac. "Instinct has to express itself somehow. Much better this way. That animal that took a bite out of your shoulder... It's what I would be without Deka. We met as children. All Relians do. We grew up together, and Deka helped me tame my old ways. He taught me how to deal with my aggression. It's why Deka won't allow me to have children. I revert too easily, and we want to keep this trait out of the population. The hope is someday we'll be free of our old ways, but that might not be for millennia."

"And the raptors couldn't just... I dunno, teach you ballroom dancing, or painting, or something? I mean, is sex really the only way?"

Kylac's ears flicked back and forth. "If it were easy..."

Deka continued the thought. "You're used to primitive urges being in the background, Stephen. Part of the subconscious. It works differently for Relian canines. Their impulses are much closer to the surface. All species are affected by primitive nature, but usually they become subdued as the conscious mind takes over. Relian canine instincts were so strong the conscious mind would never have developed had the reptiles not stepped in. If there is a better way to tame a fox, nobody has found it, and raptors have been searching for centuries."

Stephen looked down at his lap. "Then... Uh... I don't know if this is too personal, but..."

"It's your last chance to ask," said Deka.

"Right." Stephen turned to Deka. "You two act married, you talk married, you're inseparable, so why does Kylac have sex with everyone in the universe except you?"

Deka smiled with his hands. "You are not the only one to wonder that. Every raptor pairs with a fox, but it's not the kind of connection you're used to. You are used to there being only one kind of bond, a mating bond. Relians also have a different kind of connection."

"Think of it as a missing mental component," the canine added. It's instinct for raptors and foxes to take a companion for mental completeness, unlike a mating bond, which is more emotional. This is why Deka had me as a fox, and also Sonjaa as a mate. I've only known a couple raptors who thought of their foxes as both a mental partner and a physical one."

Deka clicked his hand-claws together. "Ratash."

Kylac's tail wagged. "Ratash and Irus. I wish you could meet them. They have that relationship. Don't know how they manage it. I never thought of Deka that way. I really hope those two are all right. I wonder if they were still on Jemum."

Deka continued. "The basic social unit in Relian culture is not husband and wife. It's raptor and fox. The mating bond comes second. I was close to Sonjaa, but she spent more of her time with Rupi. We met frequently, but from what I know of your culture, you'd consider it a weekend marriage. This is not instinct. We live this way because of our foxes."

"What do you mean?"

"It goes back to primitive times. The reptiles were so closely tied to their family relationships they hardly ever lived for anything else. The foxes taught them how to break away from their families and explore. When Relians discovered portals, it was the canines who urged the raptors to leave Rel. So this distance between the mated pairs, it's actually helpful. Sonjaa meant a lot to me, and I adored the hatchlings, but she was so much closer to her fox, and that's how it should be."

Stephen nodded, attention wandering down to the ground. "I never would have thought. Sorry I didn't listen earlier when you tried to explain."

"You were angry," Kylac said. "I don't blame you, and I am sorry. If disharmony is a problem, I hope Ogu can find a way to help those people."

The icy wind blew in through the entrance. A portal opened inside the cave that led to a beautiful valley full of sunshine and flowers. Deka and Kylac stood.

"Wait for Norh," Deka said. "We'll be back shortly."

They walked through the portal. It looked so warm over there Stephen wanted to leave, but he adjusted to the cold. He tried to think of other questions to ask, but nothing else came to mind.

2

Stephen felt like a sixth-grader waiting for the bus only to discover school had been canceled hours ago. Stephen touched his shoulder gingerly. It stung, but it was healing. He was impressed the Uiv had been able to patch him up so well without a doctor, or even bandages. He was also impressed Deka and Kylac had healed with very little medical attention.

Licking... Stephen wondered why his own saliva didn't kill bacteria and form a scab over the wound.

A gust of freezing wind blew through the cave mouth. Stephen huddled into himself and held his breath until it passed. He had peeked outside a while ago and thought he was back home. He had expected to see people with snow blowers clearing their driveways. It was hard to remember this was an alien planet.

Nothing to do while waiting, and Stephen didn't want to go to that warm valley until he had to. He was determined to be here when Norh came back, so Stephen sang.

He'd been told he had a decent singing voice as long as the song was in his range. He covered a few Michael Jackson hits, some Tina Turner, some Spin Doctors, and then a medley of Beatles songs. His mind wandered to some Mike Oldfield tunes, and then to Loreena McKennitt. He ran out of popular songs, so he switched to *Schoolhouse Rock* and various children's songs Alex had on tape. Then he remembered it would be around Christmastime back on Earth, so he started singing Christmas carols.

Halfway through the third verse of *Good King Wenceslas*, he heard the unmistakable sound of snow compressing. Something had landed outside the cave. Something huge. Stephen closed his mouth, and the footsteps came closer, but quieter as they left the snow.

A yellow head moved into the cave, connected by a long neck to a muscular body covered in tight scales. A pair of wings were folded against its flank. It smelled strange, but somehow Stephen felt warmer. The reptile head snorted, turned, and faced the human. It halted.

"Oh my God you're a fucking dragon!"

The body was as large as a city bus, not including the neck, tail, and wings, which made him even larger, and his scales reminded Stephen of the setting sun. They wrapped all around him, sometimes fading to black and then back again to a sandy yellow. The claws on his feet were as long as Stephen's arm. He had bony ridges across his head, black and red, and the two above his eyes looked like horns. His face was as stoic as Deka's, but somehow more placid and disarming. The resemblance was stunning—every fantasy drawing come to life. Stephen stood and stared. The dragon stared back at him.

Then Stephen remembered he needed to speak. He swallowed a mouthful of cold saliva. "You must be Norh. My name is Stephen Penarrow. Deka and Kylac brought me here."

At the mention of their names, the dragon seemed to soften. He walked into the cave and noticed the portal to the hub for the first time. He examined it, moved over to the platform of smooth rock, and climbed onto it. Stephen could not catch his breath, and all of a sudden he had enough energy to run back to Earth and divert a river.

Norh situated himself, seemed to merge into the rock bed, and lay with his head up, facing Stephen. The human kept his distance, never taking his eyes off Norh.

"My God... They told me I had to meet you, but they didn't say why. Wow... A dragon. You ever been to Earth? Or has any one of your kind been to Earth? I don't know how to describe where it is, but I'm sure you know it. Someone must have seen you! This can't just happen. The odds are— Damn! I can't believe you're a dragon."

Norh leaned forward slightly. "I remind you of something on your planet?"

Stephen was speechless.

Norh resumed. "Hello, Stephen. You said Deka and Kylac brought you here."

"Yes. Earth. That's where I'm from. Uncontacted species. Nobody's been offworld. I'm the first. Unless you count the guys that walked on the moon."

Stephen began to sweat under his winter gear. He grabbed the chest of his coat and puffed it in and out. "Damn... Why am I so warm?"

"Are you a scent-based species?"

"No. I mean kinda. Kylac tells me it does affect me. I just don't realize it."

"My scent is speeding up your metabolism."

"It is?"

"The Krone have this effect on many scent-based species. You should experience a surge of hormones as well. Do not do anything rash. You are still vulnerable."

"Oh... That explains a lot." Stephen pulled his hat off. It was wet with sweat. "What the hell? I've never... I've never felt like this before. It's like I just got out of the gym. Damn. I see what you mean—I feel like I can make the sun rise."

Norh leaned closer to the human, scented him from a distance. "You seem especially affected. How do you know Deka and Kylac?"

Stephen was panting and sweating. He unzipped his coat. "I can't believe I'm doing this. Oh my God. Is this a bad thing? I'm burning up."

"I do not know what is normal for you."

"Well, fuck this, I'm hot." He took off his winter coat. Then he took off his shirt, and he stood in only his long underwear and pants.

Norh leaned closer still and scented him. The closer Norh came, the hotter Stephen felt.

"You know..." Stephen said as he subconsciously began removing his long underwear shirt. "You look like a dragon where I come from. They were a mythical creature. Every culture on Earth drew them, carved statues of them. They never actually lived, but they are pretty famous. Even today."

Norh watched silently. Stephen was bare-chested and still burning up.

"They were said to be magical. They breathed fire, some controlled ice, others controlled the rain." Stephen frantically began removing his pants and long underwear. "They represented man's greed in some cultures. There were so many legends and stories of them. So seeing you is... Weird..."

Stephen stood completely naked, his clothes in a pile on the bedrock beside him. He stared at them. "How—? When did I take my clothes off? Oh my God, it's twenty below in here and I'm fucking sweating. What's going on?"

Stephen glared at Norh. The dragon was halfway out of his bed, leaning over. Stephen felt feverish.

"Are you enchanting me or something?" Stephen said. "This feels like a movie. I'm not the hapless victim before the title screen, am I, the one who wanders into the dragon's cave and gets eaten? That would suck."

Norh scented him from head to toe. After a minute, the dragon crawled back up his bed, lay still, and looked at Stephen. Stephen wasn't self-conscious anymore. He felt like he could walk on the clouds, shove buildings out of his way, run all fifteen miles to work and then back.

"Of everyone I have met," Norh said, "I have never smelled anyone react to Krone scent the way you have."

Stephen was still sweating. "This is getting scary. I could dehydrate."

"Uncontacted species. Deka and Kylac just stumbled across your world?"

"They made a way to my living room by accident. They told me what was wrong with the planet. I begged them to let me come with them, see how everyone else lives, and they let me come. This is the last stop, then I go home."

"Why did they bring you here?"

"They're tracking Rive and Friend. The two guys destroying planets. They were here. Deka and Kylac are trying to pick up their scent again. It's getting too dangerous for me to tag along, so now I have to go home."

Norh tilted his head. The action reminded Stephen of a dog, which made him smile. "Did they tell you anything about Kronia?"

"Not a thing. They wanted me to be surprised."

A large sphere opened on the other side of the cave, three times the size of the ones Deka and Kylac opened. Krone-size. The human had seen nothing but fairly uniform portals for so long it had never occurred to Stephen they could be larger.

Norh rose to all four feet and tipped his head at the new sphere. It showed a desert landscape dominated by a structure of black stone. "Follow me."

He slipped through the portal. Stephen considered grabbing his clothes, but he saw where they were going and figured he wouldn't need them. The further away from Norh Stephen was, the colder he felt. Stephen followed.

3

Kylac walked on all fours, scenting the ground down the river. Rive and Friend had been this way, probably to fill the time while they created a way to the next world.

"I can smell Rive is even more afraid than he was on Uiv."

Deka bent down. "I can't smell Rive at all. I can get a little of Friend's scent though. What's his mood?"

"He smells... He's been in a perpetual state of panic since I picked up his scent on Loam."

"Panic?"

"If he can't stop thinking about this, it's probably affecting his mind. I haven't found any places where they took a rest. Friend might not be sleeping at all."

Kylac scented a particular patch of ground long enough for Deka to catch up and scent it with him.

"Kylac! That's Krone piss!" Deka shoved him away from the spot with a foot. Kylac stumbled sideways, waving his tail. He found the trail again and resumed tracking.

"It was interesting!" the fox protested.

"The next world could be in ruins by now," Deka grumbled.

"Yeah... Just think if it is. We'll have to evacuate the survivors to Kronia."

"We'll give the Krone someone to care for at last."

"Or the survivors someone to worship."

Kylac trotted along the ground, leaving Deka behind. His Archeon sense of location and direction told him they had walked about three hundred paces from the hub. Kylac stopped and scented one particular area that didn't smell like the rest of the land. He stayed on this spot for some time.

"A portal intersected the ground here," Kylac said.

Deka caught up and scented it with him. "It doesn't smell like another planet. It smells like... Like..."

"It's another part of Kronia."

"Who did they go to see? Can you tell where it went?"

"I think I can... Yes, I think I know where they went. I'm making a way."

The fox rose to his hind legs and winced. Deka raised his head, and his nose lead him directly to a place on Kylac's stomach where a scab had broken open. Deka began licking it.

"That hurts. Deka, that hurts."

"Quit breaking them open and I won't have to clean them." Deka licked until he didn't taste blood anymore. "I'm worried about you, Kylac. That's the fifth time you've reverted since the first disaster. It's becoming easier."

"Uiv was different. Those people forced me to revert."

Deka was scenting Kylac over, parting the fur up and down his body, making sure the scabs were still sealed. "I don't want us to separate again, Kylac. Ever."

"Neither do I, but sometimes it can't be helped."

Deka had finished checking Kylac's wounds. He backed away and looked him over. "Kylac, I was worried I wouldn't be able to bring you back this time. It took so much to help you remember."

Kylac sighed. His legs trembled, and he sat down facing the river. "It felt that bad for me, too. Blood never smelled that good before. Your scent never made me so angry. I'm worried how much it's happening, and it seems to be getting stronger. I knew I was prone to reverting, but you tamed me so well I never had a problem until the disaster."

Deka sat beside him, his flank against Kylac's. "I think we have to face the possibility that the disaster took more away from you than your ability to hold portals open. On the next world we evacuate, I'm not leaving you alone. You were fine on Rebus, but I won't risk it. I promise I won't let you revert again."

Kylac leaned against the raptor. "Please don't."

A portal opened behind them, leading to another river valley. Kylac stepped through, Deka followed, and they stood on the grassland. Already Kylac was on all fours, searching the ground.

"I was right! Friend was here!"

Deka scented the air. "This is the same latitude, but a different valley. Why did they come here?"

"I think Rive was asking the same question," Kylac said, now a dozen paces away from Deka. "He smells more afraid than ever." He concentrated on a single place long enough for Deka to catch up. "Deka... I'm beginning to think Rive is afraid of his fox."

"Friend *is* destroying entire solar systems."

Kylac stood upright, faced Deka. "I don't mean that. Friend smells different. It's subtle, but I can tell."

"What's different?"

"He still smells like Friend, but... Something has changed. He's not the same person he was, and I don't think Rive likes it. I think Rive is making all these ways because Friend can't. He's intersecting the ground on purpose."

"If he wants someone to stop Friend, why doesn't he? He can keep his fox under control!"

"I don't think he knows what to do about this. Would you?"

Deka sighed. "If it were you... No. But I know what needs to happen."

"Their next destination is in the desert."

Kylac sat down on the sandy soil and thought about the quantum connections he needed to make. Deka turned and looked out over the water.

"Kylac." He paused for some time. "The old ways. What is it like to lose your higher mind?"

Kylac didn't answer at first. He stared at his paws, at the river, at the clouds. Everywhere but Deka.

"It's... It's fear. There is no thought. It's stimulus and response. It's panic. It's power. And when I satisfy the scent anxiety, it's peace. It is the most reassuring peace I have ever felt. It's a peace I can't feel any other way, and if I think about it, I crave it again. Without you... that's all I would live for."

Deka turned to Kylac. His fox looked so vulnerable now. Deka hadn't thought of Kylac as such before the disaster. He walked over to him and sat down. They leaned on each other as the canine calculated the next way.

4

It was cooler under the stone, and Stephen felt relieved not to be feverish anymore. Now that they were out in the open, Norh's scent didn't affect Stephen's body so dramatically. In fact, he felt quite comfortable as he walked among the tents and huts.

For a while he thought he was examining some ancient Earth civilization. The tents reminded him so much of the ones ancient cultures were imagined to make. Some resembled stereotypical tepees, painted and decorated in various patterns and shapes. The resemblance ended there, as the designs and written words definitely looked alien.

They toured the entire village, Norh leading Stephen around to each hut, translating the words, explaining why this person would have written them. The entire story of Kronia could be told from any one of these villages.

Then they stopped at the centerpiece of the town: the large sculpture of a Krone. Stephen stood before it. It had no face, and its wings were only implied, but the form was unmistakable, and it towered over the town. He knelt before it and imagined an entire population worshipping this thing, pleading for mercy as the flying reptiles unknowingly hunted their companion species to extinction.

Another portal opened behind them. Stephen stood up as Norh walked through it into a valley full of flowers. Stephen guessed they were done here and ran to catch up. He touched down in the valley, and the bare ground felt good under his feet.

Another enclosure made of black stone loomed before them. Stephen ran to catch up with Norh and then walked side by side with the Krone.

Norh turned his head to look down at Stephen. "Tell me, Stephen. If the human race realized it had just destroyed its own companion species, what would happen?"

He thought. "Species are disappearing every day. We don't mourn the dodo bird, or some rain forest frog. But if we knew it was intelligent... That's a different question. We'd still have to go to work. We'd still have to make a living. Everyone's gotta eat. I don't think it would change much. Nothing seems to, no matter what happens."

"When the Krone realized what they had done, the entire species went into mourning. You do not believe yours would care about the life it destroyed?"

"I don't think anything would be enough to make all five billion people on the planet abandon their way of life and go into self-exile, the way the Krone did."

Norh said nothing. He faced forward as they entered the enclosure. Instead of cloth and skin tents, stone buildings dotted the interior. Stephen felt right at home, as they resembled buildings on Earth, and they seemed built to his scale. He paused at one of them, pushed the wooden door open, and looked inside.

Table, chairs, fireplace. Eerily like home. He turned to the floor and noticed the channels etched into the stone. He wasn't sure what they were, but they led from the table in the center of the room out to the door. Stephen guessed it meant they didn't use the fire to cook their food.

"What do you think?" said Norh. His breath was extremely hot. Stephen broke out into a sweat just standing here.

Stephen leaned in the doorframe. "If I didn't know I was on Kronia, I'd swear humans built this. Were your companions anything like me?"

"There are few depictions of them. We know they were bipedal and had fur. We know their language. We know how life was for them."

Norh turned and walked down the avenue toward the tallest structure in the distance.

"What's that?" Stephen asked.

"That is their center of worship."

"A church?"

"If you want to call it that, you may."

They passed many houses, all resembling something out of plague-era Europe, but less dingy. They reached the church. The door was already open, and large enough for Norh to fit through. Inside were rows of pews leading up to an alter, and a statue of a Krone. This one was life-sized and quite detailed except for the wings, which had been etched onto the body rather than carved in relief.

"God, it's just like being in church. Exactly like the churches back on Earth. Well, the Christian ones."

Norh was walking down the center aisle. Stephen followed, looking all around.

"It's like they designed it for you," Stephen said. "In case you ever wanted to visit."

Norh stopped at the altar. He lay before it, hanging his head as if in prayer. Stephen stopped beside him, warming up just breathing the air coming off Norh. The Krone said nothing for some time. Then Stephen knelt at the alter and folded his hands. He heard Norh shifting. Stephen turned to him. Norh was looking down on Stephen.

"Does your kind worship something?"

"Much of Earth does in one way or another. I used to. Wife and I were churchgoers, but I haven't spoken to God since Brenda died. Never went back to church, but the pastor calls me once in a while. Nice man. I think he really does care about me. He's probably really worried now that I've gone missing."

"Tell me about the one you worship."

"Which one? There are many. Before this, people worshipped the sun. They prayed to gods of the weather, hoping for a bountiful harvest. Pleaded for a mild winter, or for a marriage to produce children. You name it, it's been worshipped."

"And now?"

"Most faiths involve one God. Keeps things simple, I guess. That's a question I didn't think to ask the guys. I haven't seen anyone performing bizarre rituals to their gods. Do aliens believe in God?"

Norh turned his head to the statue. Stephen was still kneeling in prayer.

"I think your answer is in front of you, Stephen Penarrow. Our companion species worshipped something. They worshipped us. Why would they do that? Why would anyone worship anything? What do you ask for when you speak to your god?"

"Well... Back when I was talking to God, I'd ask Him to help me stretch my paycheck so I could cover all my bills. I'd ask Him to keep the tire inflated until I could get to the shop and get the nail removed and the tire patched. I'd ask to help me get home safely when the weather got bad. I asked Him to cure Brenda's cancer."

"We hunted these people," Norh said. "They could not speak to us. They could not plead to us. We never knew they were trying. They began worshipping us, also believing they could influence what they could not control. Among all species that develop some kind of religion, this is how it always begins. Only people who exist at the mercy of something they cannot control create gods."

"Yeah, I noticed. Everyone else can live in the cold. Everyone else can live in the rain, survive droughts. Not us. We get wet, we get sick. So it dries up, and we die of thirst."

"Your race is exactly the kind that would create gods. Since you are uncontacted and isolated, there is nothing on your planet to challenge those beliefs. You are unaware of how they originated, so they persist."

"Oh, we know where they came from. Kinda. Lots of people question religion."

"And yet most of the planet still believes they can influence what they cannot control by asking the forces of nature to do things for them. My kind was such a force. It did not influence us."

Stephen had been looking for some kind of gesture of laughter. He hadn't seen one yet, and he was starting to think the Krone were humorless.

"Your species has lived its entire existence at the mercy of its planet. From weather, to volcanic eruptions, to earthquakes, disease, flood, drought, rain, and so forth. Because of this, you created various gods, and stories relating to the gods, to explain things you did not understand. Superstition emerged so people could convince themselves they had some degree of control over these things.

"The Krone did not live in this way. We can fly. We have claws and teeth, and we can hunt anything. Our scales shield us from the cold and protect us from the heat. We can survive anywhere there is breathable air. We did not endure thousands of years at the mercy of our environment. Many species in the contacted universe are like this. If a volcano erupts, they can fly away and live somewhere else. If it floods, they can do likewise. They have no superstitions. They do not pray to gods because they have never known how it feels to be helpless."

Stephen faced the statue, gazing into its implied eyes. He felt small, as if he were alive at Norh's discretion, and he knew exactly how these people must have felt.

"Not long after I matured," Norh said, "I began exploring the contacted universe. It did not interest me, so I began to seek uncontacted planets with lone species. I found one on a volcanic planet that had no name. The people were covered in hard skin to protect themselves from the heat, but they still had to hide from the poison in the atmosphere, and their villages were always in danger. I observed their villages. When a volcano erupted, I stopped the lava from reaching them."

"You stopped the lava?"

"Yes."

"How?"

"Sometimes I lay in front of the flow. Other times I picked up rubble and built a barrier. Usually I dug a trench and channeled the lava away from the village. I guarded those people for years."

Stephen blinked a few times. "You used your body to block lava?"

Norh did not answer.

"It didn't hurt?"

"Very little hurts me."

"That's hard to imagine."

"I tried to be their companion species, but they had been alone too long. Instead of trying to understand me, they worshipped me."

"I wouldn't blame them."

"I couldn't do it forever. I had to return to Kronia and rest. When I returned, I discovered the villages empty. All abandoned."

"The volcanoes?"

"The volcanoes did not need to be there to destroy them. They had warred amongst themselves in my absence."

Stephen did not speak.

"In spite of everything I did, I could not save them. No one could have. I retreated to my cave for years after that, trying to figure out what I did wrong." Norh twisted his neck to face Stephen directly. "They may have worshipped me, but a god would not had failed."

Stephen shook his head. "I can't imagine how that felt. I'm just impressed you stopped a lava flow. That's unbelievable. I'd worship anything that did that."

Norh turned and faced the statue again. Stephen understood even better why they would worship the Krone. They were as close to godlike as he had ever met.

5

The portal opened, and the Relians stepped into a cold latitude. Snow lay on the ground, but the wind was calm, so it was tolerable. Deka stood against his fox to take some of his heat, and then he saw someone flying overhead. Another Krone. He stepped away from Kylac and endured the cold. She flew over their heads and to the black stone enclosure a short distance away.

Another Krone passed overhead and also swooped directly into the enclosure. Deka and Kylac looked at each other, and Kylac dropped and scented around for a while.

"No scent. The snow is fresh."

"Let's hope they went there. Everyone else is."

Another Krone descended. Deka took off running through the snow. Kylac ran by his side on all fours. The Krone in the air swooped down, landed in front of the entrance, and stood before it, watching as the two of them approached.

The Krone's scent rushed to meet them.

"Kylac!" Deka shouted to the fox running by his side. "She smells like Sonjaa!"

Kylac felt a chill penetrate his fur coat. "What does this mean? What's going on?"

"I think I know. I won't let it happen this time!"

Deka tucked his arms in and picked up speed, quickly leaving Kylac behind. Kylac ran even faster through the snow, and they reached the entrance to the stone structure at about the same time and stopped before the Krone. She lowered her head to their level, scenting them. She lingered over Deka for an unusually long time.

"Have we met?" she said at last, in Relian.

"Let me guess," Deka said. "You've had a fascination with languages all your life and you've been told you should be an Archeon but you disliked physics so you didn't pursue it."

She drew her head back and glared at him. "We *have* met. I should know your name."

Deka walked up to her and placed a hand on her leg. "I'm Deka. Archeon from Rel. This is my fox, Kylac. We've never met."

"I am Silci. I know your scent."

"Silci," Deka said, "you smell like my mate, Sonjaa. I'm not sure what's going on, but you're the third person I've met who had this scent."

She regarded him, puzzled. "I am intrigued. You seem very familiar, Deka."

Silci turned and began walking into the structure, Deka by her side, Kylac taking a place by his.

"Where is your mate?" Silci asked as they passed under the black stone roof.

"I don't know. The last time I saw her was before the first disaster. I hoped to find her on one of the planets. Kylac and I already searched the most likely places she would be, but so far we haven't found her."

"I heard about Rel, and the other planets we lost. Rive and Friend pondering the nature of time and what exists outside the universe. I don't understand exactly what it means, but I know it's serious."

Three Krone walked in front of them. It was unheard of to see even two Krone together, and Kylac looked around anxiously at the three dozen winged reptiles standing so close to one another, talking.

Around them were buildings made of ice mixed with dirt and wood, perfectly preserved for generations just as they had been when the last of the bipedal mammals went extinct. The Krone gingerly walked between the buildings. Deka, Kylac, and Silci walked down one avenue, between two rows of them.

"We're on their trail," Deka said. "They were here, and we're trying to figure out where they went next. Any world they travel to could be destroyed. We had to evacuate Loam, and there are bound to be more people in danger."

"Incredible. All this grief and destruction caused by two Relians."

Deka huffed. "The two least likely Relians to cause trouble. Please, Silci." He hesitated. "When did you hatch?"

"Two hundred and sixty years ago."

"Well, this doesn't make any sense. Why do you smell like Sonjaa? And why do you just happen to like languages?"

"And why is everyone gathering here?" Kylac interrupted.

Deka turned his head left, right. "I think the entire population of Kronia is in here. What's going on?"

"I heard something was in here for which nobody had a word to describe. I took off as soon as I heard." She fluttered her wings and raised her head, standing taller. "I figured I would know at least one word for it."

"Where?"

"On the far side of the village."

Krone packed the enclosure, and everyone was speaking to one another, as if they had suddenly decided to rebuild the old society.

"I've never seen this many Krone at once," Deka said.

"Neither have I," Silci said. "Whatever is in here must be startling."

"I'll have a look," Kylac said. "You figure out what's going on between the two of you."

He moved forward but Deka grabbed Kylac's tail, yanking him back.

"Remember what happened last time we split up."

"Deka, it's all Krone here. Nobody will force me into anything."

"That's what we thought about Uiv. Stay with me. Both of you."

Silci unfolded her wings slightly, then refolded them. "Demanding little reptile. Is he always like this?"

"Always," said Kylac, trying to wag his tail in Deka's grip

"Silci, the last two people who had your scent died. As long as you're with me, you're safe."

Now the Krone seemed concerned. "They died? How?"

"Right in front of me. A crowd separating us so I couldn't them."

Deka let go of Kylac's tail, and they began walking again. Krone crossed the street, glancing at them. Others stood still, speaking to one another. Most, however, also walked through the village to the far end. The rumble of Krone voices grew louder as they went further and further.

Silci began walking closer to Deka. The raptor noticed but said nothing. By the time they left the majority of the houses behind, they were walking as close as they could without her stepping on him. She slowed her pace, lowered her head, and held it against Deka. He growled, leaned on her, and they walked like this for some time.

"If I've never met you before," she said, "why does this feel familiar? Why do I feel like I've known you all my life?"

Deka rubbed his neck against hers.

The gathering of large reptiles came within sight, close to the far wall of the enclosure keeping the village out of the elements. Other Krone heard them coming, and a few parted to make room for them.

The Krone surrounded a dark, shattered sphere suspended less than a pace off the ground. The Relians halted and stared at it. Silci stopped as well, head still pressed against Deka's flank.

One of the Krone swung his neck around to face them. "It was discovered just a few days ago. We have been pondering it since."

Deka let go of Silci and ran toward it. From this angle it appeared to be a convex lens of darkness, so thin it disappeared when viewed edge-on. Looking into it revealed the same emptiness.

"We've seen this before," Kylac said. The Krone in attendance quieted down and observed the two creatures who had dared to approach it. "On Faii, but an entire sphere broken and suspended. This appears to be just a single section of that."

"We're not sure what it was," Deka said. "Kylac thought it could be an area of space out of synchronization with time. The one on Faii appeared after the second disaster. Nobody touch it. You'll disappear."

"No one has," said the male.

"Out of synchronization with time?" someone said.

The Krone were beginning to talk all at once. "Rive and Friend are experimenting with time. It could be a side effect."

"But why here?"

Deka and Kylac listened to the debate, not just to hear their take on it, but to witness the Krone debating in a large group. They recognized every voice, and their Archeon senses confirmed every Krone on the planet had gathered here except Norh. Kylac had let the portal to the hub go some time ago, and he hoped Kronia's Archeon had found the human.

6

Two portals opened in front of Stephen. The human's eyes widened. Both were cube-shaped, about the size of the spheres Deka and Kylac made. Two corners pointed at Stephen directly, showing him three faces of each cube. Each contained reflections of the valley, with him and Norh spread across each visible face. Stephen sat up and stared at Norh.

Norh's wings spread slightly and settled again. "You mentioned you wanted to see a nonspherical way. Yes, it is possible."

"I've never seen it done. The guys tell me it's not worth the trouble. What did they mean?"

"It took me three times longer than usual to open these, even with a distance of only half a pace. The calculations required are more complex, and spacetime is always pushing back. Spheres are simply easier to hold open. The calculations happen automatically. This connection does not."

The cubes winked away.

Stephen smiled. "Thanks for showing me."

Stephen sat by the river in the valley where the hub would have been, but now it was empty even of portals leading to different parts of the world. Norh lay about ten feet away, still in easy talking distance. They were waiting for Deka and Kylac to return. Norh smelled where the portal intersected with the ground, and Friend's scent, and knew where they were going but did not feel the need to follow.

"I'm glad this planet is all right," Stephen said to fill the silence.

"I was not aware Rive and Friend were here," said Norh. "They did not speak with me."

"Did you know them?"

"I know all the Archeons in the contacted universe. It is hard to imagine those two as the cause of anything. Rive was always lost in thought, and Friend went to sleep and woke up thinking about the mysteries of the universe."

"What mysteries? Sounds like you guys have everything figured out."

"We understand how the universe works, but that does not mean we know everything."

"Tell me some things you don't know."

Norh's wings spread, just a little, and then folded up to his body. "We do not know why the universe exists. We do not know how it came to be. We do not know why it is expanding. We do not know what is beyond it."

Stephen smiled. "Keep going."

"We do not know why atomic particles are sometimes there and sometimes not, or why the atomic world works the way it does. Nobody has been able to make a way into a singularity. We do not know why galaxies rotate the way they do. We do not know why galaxies collect in clusters. We do not know why gravity is so weak compared to the other forces, but many are pondering it."

"Seems kinda pointless, spending all your time thinking about stuff like that."

"Knowledge is a goal unto itself for a mature species. What is pointless is to ponder those ideas while people are dying."

"What do you mean?"

"The Krone contemplated these mysteries, and more, before we discovered portals. At the same time, we were killing off our companion species. What sense does it make to be more aware of the universe than what is happening to the people on your own world?"

Stephen laughed. "That's us. We sent messages of peace and cooperation into space while we were bombing the hell out of Vietnam."

Norh remained silent.

Stephen faced him, met his direct stare. "You keep talking about it as if you were there. You weren't, were you?"

"I am among the third generation of Krone to hatch after my species realized what it had done. I think about it often. All Krone do. It keeps us humble." The Krone was staring at him hard now. He began scenting Stephen. "Kylac gave you quite a bite. You saw him revert to the old ways?"

"Yeah. The last planet we were on. Not a good visit for anyone. I don't regret going there. I was uncomfortable, but now that it's over, I learned a lot."

"You have no claws or teeth for defense. How does your kind protect itself?"

"We made knives and spears, then guns and nuclear weapons. I had a knife with me when Kylac attacked. Couldn't even use it. I lost it on Uiv, just like my shirt."

"And you built villages and churches as well? All over the planet? Even in places you were not suited to live?"

"Yeah."

"Why did your species leave the habitat it was meant to live?"

"I don't know. Probably for the same reason I left my planet and came with Deka and Kylac. I had to know what else was out there. I had to see more."

Norh said nothing. A large portal opened up in front of them leading back to Norh's cave. Stephen noticed the portal Kylac kept open to the hub was gone. Norh rose, and Stephen stood with him.

"There is one last place I want to show you," Norh said. "Pick up your warm clothes. You will need them."

"You sure? You're doing a great job keeping me warm."

"Your metabolism is keeping you warm. You will need to eat soon, as your body is using a lot of energy."

Stephen took a deep breath, stepped through the sphere, and picked up the pile of fabric. He double-checked to make sure he had every layer, including the shoes and socks, then stepped through the sphere again, back into the warm valley.

"Damn, that's rough! Naked in fifteen below. Ow!"

He dropped his clothes, found the long underwear shirt and slipped it over his head. "So where are we going?"

Norh said nothing but watched Stephen closely as he rifled through the pile. He found his boxers and slipped them on.

7

Deka and Silci stayed with the group of Krone around the section of a broken sphere as they debated what this thing was, where it came from, and what it could mean. Kylac never thought he would live to know a day when the Krone gathered in one place again—it was simply not done anymore—but the entire species was here, mingling, socializing for the first time in generations.

Kylac dropped to all fours. The snow absorbed most scents and made them inert, so tracking was difficult. Still, Kylac hoped. He wandered away from his raptor, around and under many Krone, searching the area. He wished so many Krone hadn't been here.

Kylac looked back at Deka and Silci. Deka was keeping her as close as possible. They acted as if they had known each other for years. Kylac's ears flicked back and forth thinking about it, and then he continued searching.

He wandered into the village, tried a few houses, found nothing, and then wandered into the structure in the middle of all the houses. The pews here were made of wood, and the statue of the Krone was close to life-size, made of the same stone as the floor. These people had gone through a lot of trouble to find stone and carve it in the middle of this frozen wasteland.

Kylac walked down the aisle and stopped at the altar. His nose caught something, pulling him down to the ground, and he scented it deeper. Rive's single footprint. Kylac's tail wagged wildly, and he scented around and around.

His nose took him to the statue itself. He smelled fox here. Friend's scent. He had placed a single hand on the statue. Kylac tracked the scent around this frozen place of worship, up and down the aisles, every pew. He caught brief glimpses of Friend and Rive, and there didn't seem to be any purpose or direction. They were wandering.

Then Kylac's nose caught two familiar scents, and he raised his muzzle. Stephen and Norh stood in the large doorway. Kylac's tail wagged, and he rose to his hind legs.

"Stephen, you met Norh! Has he been treating you well?"

"Hi, Kylac," Stephen said, walking down the aisle between the pews. "What are you doing here?"

"Every Krone on the planet is here. We followed Rive and Friend to this village. I'm scenting for a portal intersection."

Norh walked down the aisle, up to Stephen, and stood with one leg right up against him. Unusually close, Kylac noticed, as Deka and Silci had been.

"Norh showed me a few villages. He told me what happened to these people. I'm at a loss for words."

"Everyone always is," Kylac said. "The entire contacted universe recognizes it as the saddest story of a planet's development ever told. So, tell me." Kylac's tail wagged. "What did his scent do to you?"

Stephen laughed. "Fifteen below, and I start burning up. I stripped down all the way, and I was still sweating!"

"Norh, you made him strip in that weather?"

"He smells better without them. I brought Stephen here because this is a rare event. I have already been here, but I left because I did not believe there was anything new to discuss. Gathering in one place for too long dishonors the Lost."

"Well, in this case I think it's necessary," Kylac said. "The negative lens is by the far wall. Just follow the noise."

"Is Deka there?" Norh asked.

"Yes. With Silci. Have you met her?"

"I did, long ago."

"Has she always been here? She didn't just pop in all of a sudden? She's lived here all two hundred plus years?"

"Of course. Why do you ask?"

"You better find them. Let Deka know I'm searching for Friend. He won't like it, but he's busy with Silci."

"Busy?" Stephen said.

"Just go, find them."

Kylac dropped to all fours. Stephen turned around and walked out, Norh keeping pace with him. Kylac watched them go. His ears flicked back and forth, then he resumed searching.

8

Stephen stood next to Deka, who was staring at the dark thing at the center of everyone's attention. All around him the Krone were discussing it, circling it, examining it. No one dared go too close. After several minutes, Deka noticed Stephen was staring, and he nudged the human in the side with an elbow.

Stephen's trance broke. "What?" He looked around and realized he was standing between two Krone, Norh and Silci.

"Wanted to make sure you were still awake. And you're sweating."

"I know," Stephen said, removing his hat. "It's all these dragons in a closed space. It's making me hot."

Deka held his claws together. "I wondered how Krone scent would affect you."

"Why does it make my heart race?"

"The Krone have a similar effect on everyone who responds to scent. Makes you want to run orbits around entire solar systems, doesn't it?"

"I feel like I can. Jeez." Stephen was removing his gloves and stuffing them into one of the coat pockets. He gestured to the sphere fragment in front of them. "What is that?"

Deka was about to speak, but Silci turned her head and faced the human. "We don't know, but we're listening to the Krone discuss it. Maybe they'll come up with an answer. Take this in, Stephen. You will never see this many Krone together at once in your lifetime."

"You speak English, too? You know... You remind me of someone."

"Me, too," Deka said.

Silci brought her head down and held it against Deka. Deka leaned on her. Stephen stared at them for a moment, then turned to Norh, who was also staring at them. Around them, the Krone debated.

Silci rubbed her head against Deka's flank. She was growling. She began to sound the same as Sonjaa, and Deka held her.

Silci switched to Relian. "Follow me."

She turned and walked toward the village. Deka followed her, glancing back at Stephen and Norh. The Krone stood close to the human. Closer than Deka himself had ever dared to stand.

Deka followed Silci all the way to the temple at the center of the village, and they descended the aisle, passing each row of cold pews.

"Who is Rupi?" Silci asked.

Deka stopped and stared up at her. It was disconcerting only to be able to see the underside of her chin from this angle. After a moment, she turned her head away from the statue and looked down on Deka.

"Rupi was Sonjaa's fox."

"I kept thinking about her while they were debating. The name wouldn't leave my mind. Have I met her?"

"No. No, Silci, you have never met her. Sonjaa took Rupi everywhere."

Silci lowered her head to Deka's eye level. Deka resented her size. It prevented him from being closer to her.

"Then why can't I stop thinking about her? Wondering where she is, how she's doing. She was never more than a reach away. She was calm. She was quiet. She shared Sonjaa's enthusiasm for language, but she enjoyed language as music."

"That's right."

"Rupi was... She was carrying a child."

"Yes!"

"The father was... Grei?"

"Gren."

"That's it. Gren. Gren and Rupi. I thought about them often. Always those two. Either them, or you and another fox. I remember all of this. All my life I've had little flashes of it. People I thought I knew. Places I never visited. Over and over I thought about Rupi and Gren. Rupi was finally pregnant. Gren and his raptor. I remember speaking to Gren's raptor, telling him Gren and Rupi would be a good match. When the father smelled his child, he stayed with her. Beautiful to watch that happen in Relian canines, how potent paternal instinct is when a father smells an unborn child is his. They become intensely loyal and caring. Now they're going to raise a child at last. Whenever I let myself think about it, I was so happy. That's as far as the feeling ever went."

Deka was speechless for several breaths. "Did you ever come to Rel?" he said at last.

"Not recently. I never found anything that reminded me of what I knew. Nobody knew anyone by those names, so I haven't been back in over a hundred years."

"And you've felt this all your life?"

"It's been getting stronger since the disaster. Meeting you brought everything out again."

Deka stepped forward, held her face with one hand. "Silci, please listen to me. I think your life is in danger. Something bad is about to happen. I don't know what, but if we're separated, you will die."

"What could possibly kill a Krone?"

"That lens, for one. It could definitely kill someone. There are a lot of Krone here. It's similar to what happened on the other two worlds. It's about to happen again. Please stay close to me."

Silci regarded him, took in his scent. From the way her body moved, she felt the same way about Deka that he felt about her. It was a great comfort. "I promise."

"Speaking of people who shouldn't leave my side," Deka turned his head to the statue. Kylac was standing on his hind legs, leaning against it.

"I have good news," said the fox, also in Relian. He pushed away from the statue and walked up the aisle toward them. "I found another place where a portal intersected the ground. It's behind the statue. Some of the stone floor is now dirt from Walin. I'm working on a way right now."

"Good find, but I still don't want you to go too far away from me."

Kylac sat on one of the pews and faced both of them. "You have more than enough to watch over." His tail wagged. "An entire Krone. Silci, you do smell like Sonjaa. It's not just Deka who thinks so. Something strange is going on."

"You smell familiar, too," she said. "The feeling is stronger. I've known you all my life, but I don't know how."

"We should leave," Deka said. "Where do you live? I'll make a way for us back there."

"I can't leave now. I want to know what that thing in the village is."

"They can debate all year and they'll never figure it out. Please, let's get away from it. We can go anywhere you want."

She growled a little, raised her head and looked down at Deka. "I will be part of the debate."

She turned and began walking to the exit.

Kylac stood and bumped noses with Deka. "So much like Sonjaa."

"Silci, please don't go back there."

She left the temple. Deka ran after her. When she noticed he was following, she took off running, wings outstretched.

The three of them rejoined the Krone surrounding the lens. Stephen, now completely disrobed and still sweating, was dangerously close to it, mesmerized, walking around it to examine it from every angle. Norh watched him closely.

Two other Krone were talking amongst themselves about how such a structure could form, discussing the mathematics and physics needed to calculate such a thing, and how such calculations might scar spacetime, leaving this behind. Norh did not participate. He stayed just behind Stephen, watching his every move.

Kylac wagged his tail. Deka rubbed his claws.

9

Night had fallen, and the Krone were asleep. Cold and heat did not affect them, so they slept directly on the packed snow. Stephen and Kylac slept on Norh's belly, Stephen still naked but not sweating anymore.

Footprints, a Krone's and a theropod's, led to the entrance. Outside the enclosure, the wind howled, and the snow blew into drifts twice as tall as a Krone against the black stone. Silci lay just outside it, blocking the wind and snow with her wings. Deka leaned against her shoulder, not ashamed to benefit from the heat she gave off.

"Sonjaa met you nine years after she met Rupi," Silci said. "When she caught your scent... She just knew it was you. It had to be you. You were perfect. You smelled it, too, because you never left her. For those first few years, only Kylac could pull you away from her."

"How did we meet?" Deka said. "Do you know?"

"I think so. It was during your training as an apprentice. Your Archeon was helping you learn to make a way to a different part of Rel, but you accidentally opened a way into the ground. Sonjaa fell in and spilled out in front of you. Kylac was rolling laughing. She was furious, and you were the one to calm her down. And your scent... It was more than enough to calm her down."

She was silent for a moment. Deka rubbed against her. Hearing her say these things brought back good memories. It had been so long since he thought of those days.

"Like now," she continued. "Something about your scent is comforting. My mind is chaos all the time, but you seem to calm it. It's like being pulled apart. Back then, Kylac was always pulling you away, and I was pulling you back to me. Same for Rupi. She was always pulling me toward her, and I was trying to pull all of us together. It was strange. I never thought I would become so involved so fast.

So quickly after I met you, I wanted to have your eggs. I knew what was happening, but I didn't want to stop it. I was so taken by your—"

Silci stopped speaking for a moment. Deka leaned against her harder. She smelled more and more like Sonjaa with every breath Deka took.

"I mean, *she* was so taken by your scent. Deka, is this how it feels to desire a companion?"

"Yes," Deka said, breathlessly feeling her scales with his neck.

"I heard about it from my time on Rel, how intense it is for the reptile species. A primal drive to take a mate. The drive to nurture a fox is even stronger, and whenever I think of Rupi, I can't rest until I find her. Then I realize she doesn't exist. All my life, Deka, I've thought about things like this. Little flashes of memory when I'm not thinking about anything else. When I was young, I would ride these thoughts for days, but they were always so vague and disconnected until you arrived."

Deka growled softly. "Come with us."

"To where?"

"Kylac is making a way to Walin. It was Friend's next stop."

Silci did not answer right away. Deka continued rubbing against her, even more frustrated by her size.

"Deka, I am not your mate."

"You are."

Silci rose to her feet and stepped away from Deka. The raptor almost fell over, but he caught himself and remained standing, gazing up at her.

"Silci, I'm an Archeon. I am aware of everything my brain takes in. I am sure you are Sonjaa."

"Deka, I've never met you until today."

"But you remember Sonjaa's life as if you lived it. You're having a difficult time separating your memories from hers."

Silci leaned away from him.

"The memories are getting stronger," Deka continued. "Aren't they?"

"I'm a Krone. I do not desire a mate, and I certainly do not want a fox. I'm still another century away from estrus. When the portals return, I will continue my travels to other worlds, learning the languages. I don't need a Relian following me everywhere."

Deka was hurt, but he also remembered he was speaking to a Krone. "Then... tell me just one more thing. Do you remember where you were when the disaster happened?"

"No. I didn't know there was a disaster until I flew to the hub and found it empty."

Deka could think of nothing else to say. The icy wind blew around them. Deka shivered a little.

"Although..." she stared at the ground. "I remember being in a group of Relians. It was Rel. Yes. I was on Rel. Or—no, Sonjaa. *Sonjaa* was on Rel. She was with another fox. He was doing something. Opening a sphere. I ran to him, trying to stop him. And... That's all I can remember." She looked up at Deka again. "I try not to think about that one. It always leaves me with this strange feeling."

Deka had been holding his breath. He finally let it out. "A group. She was there when it happened?"

"You were there."

Deka turned to face her.

"I remember seeing you. I couldn't smell you, but I recognized your scale pattern. You saw her... you saw her die."

"I was stabilizing the portal to another world. I was on the other side of the planet."

"No," Silci said. "You were there. I remember. Deka..."

The raptor did not like the change in her scent. He stepped closer to her.

"Deka! I know what that thing is in there!" She turned her head and gazed into the entrance. "They're right. It *is* an empty area of time. And I—"

She froze. The wind howled and snow blew around them.

"Sonjaa, what's wrong?"

"It's Friend. We told him to stop thinking about time. We warned him. That's not a sphere! I can stop him!"

She spread her wings, flapped twice, kicking up a tremendous amount of snow, blowing Deka to the ground. The snow covered him, and he struggled to stand. By the time he climbed out of the drift, Silci was flying over the village.

"Sonjaa!" Deka screamed, already dashing at full speed. "Come back!"

Houses sped by. He dashed past the temple and down another street, straight for the negative lens.

"I can stop him!" she shouted. She began to descend, aiming for the broken sphere.

Deka ran toward it. It shined its emptiness directly at him, it seemed. As he approached, he heard raptors and foxes, smelled their scents all around him.

An enormous group of sleeping Krone blocked his way. Deka growled and leaped, landing on a Krone's back. Silci was flying straight for the lens. Her scales had changed to brown and green, and her snout was much more like a Relian.

"Sonjaa, stop! Stop!"

Silci careened toward the broken sphere. "Friend, close it! Close it! Now!"

Deka ran along the Krone's back, leaping from one to the next. As he landed on one male's back, every Krone in the cavern stood, and Deka tipped over, rolled down his flank, and crashed to the ground. He rocked himself back up to his feet and faced the lens, but a forest of legs blocked his way as the Krone woke and began running aimlessly.

Through their legs, Deka saw a Krone with Sonjaa's scale pattern dive straight into the shattered lens. She vanished into it, snout, wings, legs, and tail.

"*Deka, help me!*" Sonjaa's voice came from the broken sphere of spacetime.

The raptor screeched and leaped but couldn't clear the stampeding reptiles.

"*Deka, what's happening?!*"

The raptor leaped up again and again. Instead of Krone, he smelled Relians everywhere. Finally he sank his claws into a Krone, climbed up his flank, and stood on his back. His claws didn't even pierce the Krone's scales. Deka stared at the lens. It remained as empty as ever.

"*Friend! Rive! Help me, please! Someone! Help!*"

Deka stood still as the Krone searched for the source of the voice.

10

Kylac sat with Deka by the river where the hub should have been. The calm murmur of the water was the only sound. Deka was shivering despite the warm breeze coming through the valley, the fox draped around him, and the Krone scent that should have sped up his metabolism. Every now and then, he chirped in pain. Stephen sat on a nearby rock, still undressed. Norh lay near him, neck between the human and the Relians, also watching the river.

"Walin..."

It was the first word Deka had spoken all day.

"I'm a couple days away from finishing the way," Kylac said.

"Good. Because I lost the way to Earth."

"It's all right. We'll send him back from Walin. There's no hurry."

Deka's spoke through clenched teeth. "When I find Friend, I'm going to kill him. Rive won't do it, but he wants someone to. I volunteer."

"Think, Deka. Don't do anything stupid."

Deka started to scream, but it degenerated into a whimper. Kylac held him for a while, and then he turned to the Krone.

"Norh, if I tell you how to get to Earth, will you take Stephen home?"

The Krone turned his head and faced the fox. "That won't be necessary. I'm coming with you. I will care for Stephen."

Kylac's ears bloomed. Stephen looked up at him, too.

"Are you sure?" Kylac asked. "We're sending Stephen home because this is getting too dangerous."

"You worry about that. I will worry about Stephen. But please tell me where Earth is."

Kylac spat out relative position, velocity, orbit shape, vertical oscillation, rotational speed, tilt, axis wobble, planet shape, and a whole slew of other information. The whole time Stephen was looking up at the Krone, not understanding a word. Finally Norh turned to the human and regarded him in silence for some time. His wings spread slightly, then folded back against his sides.

"This is how my kind smiles," said Norh.

Stephen smiled back. "Thanks. But why? You barely know me."

"I know your scent. If you want to see more of the contacted universe, I will take you."

Stephen glanced at Kylac. The fox smiled like a human and nodded vigorously.

Stephen met Norh's eyes again. "Yes. Thank you."

"We will follow Deka and Kylac for now. When you are ready to leave them, we will go alone."

Stephen nodded. "All right. I'm game for that. I'd love to see more."

Deka spoke. "It'll be good to have someone who can keep a portal open longer than a few breaths. You might save our lives, Norh."

Kylac rested his head on Deka's. They listened to the river. Stephen couldn't take his eyes off Norh.

S'rin

The portal opened over the grassland, air rushing around it. A Krone emerged through the sphere large enough to accommodate someone his size. Stephen walked through a moment later, wearing pants and his long underwear top, winter clothes stuffed in the backpack. Deka and Kylac walked through together, and the four of them surveyed the planet.

The wind was powerful enough the bipeds held on to each other to keep from being swept away. The Krone stood as though nothing were different.

This solar system's star appeared to be untouched, but a jagged horizon half-surrounded them. Pieces of the planet floated in the distance, where the world had been cut away. At least a third of the planet was missing.

Norh walked toward the fallen horizon. Stephen followed him, fighting the wind, just barely managing to stay upright. The Krone stopped at the ledge and peered down. Stephen held one of his legs and looked down with him. Another diagram cutaway of the interior layers of a planet. The magma had long since spilled out, and everything below the crust was cold.

To look out over the cliff was to look out into space itself. Chunks of rock, cooling magma, and lifeless bodies floated everywhere. Stephen swallowed. As many times as he'd seen it over these weeks, it surprised him every time.

Kylac was on all fours scenting the ground, Deka following a few paces behind him. The wind was too strong to hear one another easily, but Deka and Kylac silently decided to make ways to the different regions to search for survivors. They estimated they had four more days before the atmosphere was gone, plenty of time to evacuate the people through Norh's sphere.

Kylac concentrated on a particular spot. Deka caught up and scented it with him. He didn't have to be a fox to determine this piece of soil didn't belong on S'rin. Their eyes met. They both knew where Friend had gone. They stood up together. Kylac shouted at Norh.

"Norh, begin thinking about Hesiv," Kylac shouted when Norh was in earshot.

"I will make the way."

They huddled in front of the portal, Deka holding Kylac and Stephen. Kylac and Stephen held the bone ridges on Norh's head as the wind rushed off the planet and into space.

"Deka and I will make a sphere to each side of the continent. There will be survivors, and they'll be scared."

"Who lives here?" Stephen said.

"This is S'rin," Deka said. "I don't think the Teshi have a parallel species on your world, Stephen. Their companion species is the Tesha. They'll be more familiar. They're rodents, similar to the ones you saw on Rebus."

"I can handle rodents. And two-headed snakes, and six-armed dogs. Easy. Just warn me before we meet another insect species. My ass still hurts."

Deka couldn't laugh, so he rubbed Stephen's fingernails instead. "Sorry, Stephen, I thought Norh was going to keep you safe."

"It was perfectly safe," said the Krone, wings unfolding and folding again. "You have not lived until your digestive tract has had a complete cleaning from the Estons."

"Yeah, thanks! Next you'll take me somewhere to get a free dialysis! My eyeball fluid is due for a change, too. Who can do that?"

"The Selts already fixed your eyes!" Kylac said.

The wind gusted. They braced themselves until it calmed enough to hear one another again.

"The first time we came here," Kylac said, "I knew nothing about these people. The Teshi and the Tesha communicate by sharing brainwaves."

"They're telepathic?" said the human. "So we have to watch what we think?"

"Doesn't work like that," Deka shouted. "They can't share thoughts with any other species. Our brains are not compatible."

"So we have to get by with just words then?"

"Telepathic species do not have a spoken language. It never occurred to them to develop one."

"So how do you talk to them?"

"They have a written language so offworlders can understand them. It's actually the only way."

"When we first came here," Kylac said, "we weren't even Archeons. I heard about this planet, and I thought the same as you, Stephen. I wanted to share brainwaves with someone! So I dragged Deka clear across Rel to find the portal that led to S'rin. There's only one settlement, but it's spread out over the whole grassland. It stretched that way, too, and we walked out there." He pointed with his muzzle over the horizon.

"Kylac and I waited for someone to talk to us," Deka shouted. "It was quiet. These people don't even comprehend sound. They are deaf. So we're screaming at them to say something, talk to us. Nobody notices we're there."

Kylac picked up the story. "And then one of them stopped us. She was carrying a tablet. She wrote some things, but we didn't understand. We were just children, and Rel has no written language, so she kept us here until we learned some of theirs."

"Didn't your parents worry about you?" Stephen said.

"We didn't need parents by then," Deka said.

"Oh. So how long were you here?"

"Nine days by your measurements," Kylac said. "She kept us until we learned enough to know what it was like to be telepathic. We had to settle for text on clay tablets."

"You learned an entire language in a week even before you were Archeons?"

"It was actually the first time someone told us we could have the ability to be Archeons," said the raptor.

"But talking with text only?" Stephen continued. "Sounds tedious and uh... What's the word I'm looking for? Hollow, distant...?"

"Impersonal," Norh said.

"That's it. Thanks. Jeez, an alien knows English better than I do."

Norh's wings unfolded and flapped a little.

"For offworlders, this place is impersonal," Deka said, "but try to imagine being aware of everyone's thoughts all the time. They don't have a sense of self-identity. You thought the Uiv were disturbing? This village is a collective mind made up of two different species. They learned how to interpret each other's brainwaves, and now they're practically one species and one mind."

"But their mental link depends on the others nearby," Kylac said. "If nobody is within range, there will be a gap in their connection with the community. With so many deaths, people will be isolated and scared. We'll have to reach them separately."

"How far is their range?" Stephen said.

"About ten paces."

"That's it? That's, what, twenty feet? That's as far away as they can be from each other before they lose touch?"

Deka rubbed Stephen's fingernails again. "If it makes you feel any better, Stephen, there's a species of telepathic fish that has only half a pace range."

"And some require direct contact," said Kylac. "Think about that. You have to touch someone to communicate. Relatively speaking, these are strong telepaths. Be aware, Stephen, their minds give off electromagnetic signals. It can be disturbing if you don't learn how to shut it out."

"What signals?"

"Noise. Remember Eiae?"

Stephen nodded.

"Remember how we told you electromagnetic waves affect people without them even realizing it? It can happen here, too, because these are people's thoughts. The mind is aware of some kind of input, but it can't interpret it, so it's just noise. If you start to feel uncomfortable, go outside their range for a few minutes."

The human nodded.

Two large portals opened to either side of them, both just a few paces away from the large offworld sphere.

"Kylac and I will take the first region. You and Stephen take the other. Bring as many people back as you can. Tell everyone you meet what's happening. We won't have any help this time. It's just us."

They broke out of their huddle. Deka and Kylac ran into one. Norh and Stephen passed through the other.

Through the portal to the planet He'Kri, birds the size of Stephen gazed at the destroyed planet, six different species, all with garish plumage and elaborate feathers. Some had a mixture of scales and feathers. They could tell the wind was exceptionally strong, and they dared not venture into it.

2

Stephen and Norh landed in the middle of a vast, open field of grass. The wind felt weak here, gentle but constant, bending the grass forest in one direction all the time as the prevailing breeze hurtled off the planet never to return. It smelled of death and pollen. Stephen was drawn to the grass, which stood twice as tall as he was. He placed his palm on one of the blades. It felt like a simple blade of grass, but it behaved more like a tree in how unyielding it was to Stephen's touch. It felt sharp, too. Stephen ran his hand down one blade but stopped because it was about to slice his hand open.

Norh turned his head and faced Stephen. "Their entire population lives in this grass forest."

Stephen just now noticed they were standing in a wide path cutting through a forest that was all leaf and no bark. Norh walked down the path, which was just wide enough for him. Stephen walked behind him, looking back and forth at the blades on either side.

"All of this is about to die," Stephen said.

"We are here to prevent the culture from dying as well."

Norh stopped, scented the ground, and then stepped over several dead bodies of vaguely rodent-like creatures. Things covered in insect-like skin and fine, green fur lay among them.

"My God..."

Norh was still walking. Bodies covered in various shades of green fur lay everywhere on the path as if they had just dropped dead.

Norh turned around, his head and tail whipping through the grass. Stephen expected to see blood—the grass was so sharp it would cut anything—but when Norh swung his head around and lowered it to Stephen's eye level, he didn't see a scratch on him.

"I will fly and find survivors. You are to make these symbols any way you can."

He extended a claw and made lines in the dirt. They looked like three variations of the number three. The first had two dots on either side of it. The second had three dots and an extra long tail. The third had an extended top line and a curved mark striking through the center of the digit.

"What's it say?"

He pointed to the first. "This is their symbol for Relians. This is their symbol for help combined with portal and the verb to go. The last is their symbol for planet, destroyed, and flee to this location joined into one. In your language, loosely, it reads Relians are here to help you. Your planet is destroyed. Go to the portal and flee."

"All of that in three symbols?"

"Their written language is full of contractions. This character," which resembled Stephen's numeral three, "is the base of their grammar. Everything around it is a modifier. An entire sentence can be compressed into one symbol. It represents how their minds work. They do not think in sentences. They share thoughts instantly, so there is no flow of information. Everything simply is, all at once."

"I think I get it."

Stephen drew the symbols on the ground over and over, sometimes with his eyes closed, hoping it would help him remember faster. The blowing wind reminded Stephen time was running out, and it also blew Norh's scent over him. Stephen didn't burn up every time he was alone with Norh anymore—his body was used to it now—

but he still wanted to climb mountains, break barriers, conquer continents, and above all not be a disappointment.

On the last couple of planets they had evacuated, there had been little for him to do but help carry people off the decimated worlds. This was the first time Norh had asked him to do something specific.

Stephen had witnessed Norh in the blistering heat carrying a dozen people in his claws to safety. Stephen, Kylac, and Deka had been incapacitated and almost passed out from heat stroke.

On another world, Norh ran through a frozen carbon dioxide snowstorm, pulling people from the avalanche that had covered the entire region in the wake of their disaster. It had been too cold even for Kylac, but Norh dug through it as if it had been sand, pulling people out by the dozen and carrying them through the offworld sphere.

Another planet had been covered in boiling lava as a result of the disaster, and Norh had shielded hundreds of people from the flow with his own body. When he emerged, there wasn't a single mark on his scales, and all Stephen could do was stare at him in awe. He had never seen something that brought him to his knees before. He hadn't even been conscious of dropping to his knees, and he believed that must be what a true spiritual experience did to a person.

Nothing touched the Krone, and he could do anything, go anywhere. The longer Stephen was with Norh, the more helpless he felt. Stephen couldn't even do the most basic things, such as hunt for himself, or figure out which fruits and vegetables he could eat. Norh did all of that for him, and the Krone seemed eager to do so, which only made Stephen feel worse. He was starting to piss himself off with how inadequate he was compared to this Krone, and Stephen was determined to impress Norh somehow.

"Will you remember?" the Krone said after the twentieth time Stephen wrote the symbols.

Stephen stood, embarrassed he needed so much time to remember the simplest things. "I have it."

Norh spread his wings. "These people will have felt the death of thousands of their own. The effect is similar to what happened to the Archeons after the disasters. Their minds were ripped away from this universe, and it killed these people. It has likely made others unconscious. The survivors do not know what happened, and they will be isolated and scared."

He flapped his wings. Stephen braced himself, standing against the gust of wind the Krone kicked up as he took to the air. Stephen followed him down the path. He wondered who could have cut this path, let alone how anyone could live in it.

After several bends and curves, Stephen came to a place that resembled a traffic circle. It was about eighty feet across, perfectly circular, and branched off into a dozen spokes. Many corpses lay here, barely decayed, but they still reeked. Stephen covered his nose as he scanned the clearing. So many bodies piled one atop the other left barely any room to walk. These people had collapsed where they stood when the antisphere hit.

Stephen looked up. Norh was circling something in the distance, directly over one of the spokes. Stephen ran down the spoke that seemed to lead toward the Krone. The path narrowed until it was just wide enough for two people to walk abreast.

The path ended at a grass hut. The grass itself had been bent over, twisted together, and attached with some kind of clear glue. Stephen stooped and ducked under the low opening. It was warm in here, and Stephen guessed it was waterproof as well. Fifteen people huddled inside the hut, more bodies stacked against the far walls. Stephen tried not to look at those and focused instead on the survivors.

The rodents were about Stephen's size, all green in color, just like the grass. They lacked ears but still had a nose, eyes, and a snout resembling a rabbit's.

The other people were much stranger. They had white skin, thin fur of varying shaes of green, but no male or female parts.

Stephen walked closer, holding his hands up out of habit, forgetting that many species considered that a threatening gesture.

"Hello?"

Nobody reacted to him. He was within ten feet, and they still hadn't noticed him. That's when Stephen remembered they were deaf.

He approached one of the rabbit-like things and touched it on the shoulder. It turned and glared at him, nose twitching. Stephen could see it was female, she had no whiskers, and her eyes were black and empty. Her lack of ears was the most disturbing part of her body. Stephen couldn't help but stare at her chest. Three rows of breasts. It looked bizarre. Now that Stephen saw one up close he realized the body reminded him of a rat, but the face was vaguely rabbit-like. The others stood in unison and stared ahead blankly as the one rodent-thing scented Stephen over.

Stephen noticed the males had bigger balls than the rats on Rebus.

The non-rodent species did not resemble anything Stephen had ever seen before. Under their thin fur seemed to be a thick shell, reminding Stephen of an exoskeleton, as if an insect and a marsupial had crossbred.

The rabbits had very large teeth, also green. Stephen guessed they were the ones who cut the grass. The other species, however, had no teeth. One of them had its mouth open, and the saliva reminded him of mucus.

Cutters and gluers. Stephen guessed they were both prey species who sought shelter in the grass to escape predators, and survival meant cooperation. Stephen felt proud of himself for figuring that out. Hearing so many stories of each species taught him to think this way, and he hoped he would have a chance to ask Norh about it soon.

When the rabbit-thing had finished sniffing him, Stephen knelt, and he scratched the three symbols into the soil. To his relief, he didn't forget the details. Norh's scent made him want to remember, and he had committed it to muscle memory.

He stepped away from it. Two of the nonrabbit-things approached the words. After a half-second pause, suddenly all fifteen people perked up at the same time and bolted out of the shelter.

Stephen ran after them. He followed everyone to the wheel, where they fanned out into the spokes. Stephen waited for a moment, checking the sky, but did not see Norh anywhere. His heart raced for several minutes, waiting, waiting as the wind blew.

Suddenly Stephen felt hundreds of footsteps. Rodents and nonrodents stampeded from every spoke, down the center, out the spoke Stephen had come from, and vanished. Half a minute later, Norh landed beside Stephen.

"That was easy," Stephen said.

"We were fortunate they weren't too far apart, and they knew where everyone else was. They were simply scared to cross the region where there was no thought."

"So let's find the rest of them."

Norh spread his wings, took off, and soared onward. Stephen followed him down the spoke he flew over. Norh rose higher in the air, circling something in the distance. After only a few dozen feet, Stephen came to another circle with spokes branching off of it. Norh circled at the end of one, and Stephen ran through that spoke and came to a sudden stop at another hut.

He crawled under the narrow opening and stood at full height. A hundred people huddled together in silence. If Stephen hadn't known they were sharing brainwaves, he would have assumed they were praying.

This time one of the insect-like people saw him enter. Quickly Stephen dropped to the ground and drew the symbols. The furred insect ran closer on its backwards legs and aimed its face down at the symbols. Everyone in the room perked up and ran to the exit in unison. Some paused to scent Stephen. Two of the nonrabbit-things ran to the far side of the hut, grabbed a couple clay tablets off the ground, and approached Stephen with them.

They stopped before him, chewed on the tablets, clawed them, and then handed them to Stephen. He took the two tablets and held them up. The people bolted for the exit without waiting for Stephen to answer.

They had pressed eight symbols into each tablet, all of them variations of the number three. To Stephen's eyes, it looked like 3 3 3 3 3 3 3 3 on one tablet, and 3 3 3 3 3 3 3 3 on the other, with all sorts of dots and dashes and underlines and strikethroughs and hooks and tails on and around each digit.

Stephen pushed his thumb into the clay. Dry, but still pliable. He carried the tablets with him as he rushed out the door and to the wheel. He stopped there and looked up but couldn't see or hear Norh at all. Stephen squinted through the dim sunlight and searched the sky.

The Krone swooped over the wheel. Stephen closed his eyes against the gust of wind, and when he opened them again he saw Norh flying over one of the spokes. He ran after him, keeping the tablets in a tight grip between both hands.

3

Deka and Kylac jumped over the bodies as they made their way down the path through the grass.

"Can we save the grass?" Kylac asked as they climbed over a bunch of Teshi, all dead.

"Who cares about the grass!" Deka shouted.

"It's such a unique plant. It's wrong for it to die. We should ask those birds if they'll try it."

"It will overtake the entire temperate region."

"The birds on He'Kri live in the trees. They almost never touch the ground. Seems like a perfect arrangement."

"The grass will choke off the trees. Then the birds will have to risk getting torn to pieces if they land anywhere."

"There has to be a way. The S'rin won't want to live out in the open. They need the grass."

"They'll have trees to live under instead. Just as confining, but only half as lethal."

"What will the Tesha do without grass to cut? What will the Teshi do without grass to shape? Maybe they can keep it under control—"

"Kylac, just focus on finding the people and getting them out of here! We may only have a day before the air is gone!"

"I don't want to rescue them so they can live in misery. It won't be home without the grass."

Finally they came to a clearing and paused, scenting the air. Death and decay were everywhere, but they detected living scents in the distance, and Kylac led the way down one of the paths .

"They can move to a place where the grass won't be as invasive," Kylac went on. "I can think of... Three planets where it might assimilate well."

"And what about the people already living there? Will they like grass sharp enough to cut bone planted on their world?"

"I'm sure they will come to find the beauty in it. How this plant evolved to survive, and the people evolved to live in it to escape the predators. Someone will understand."

"Focus, Kylac, focus."

Finally they came to a hut. They crawled under the entrance and rose to full height. The group of Teshi and Tesha recognized the offworlders and two broke from the group and brought them one tablet each. Deka and Kylac wrote symbols and showed them to one person. They came to life and dashed out, reeking of trepidation, but on Deka and Kylac's word alone, they ventured into the open, devoid of the thoughts of others.

Deka glanced at Kylac's tablet.

"Kylac, is now really a good time to make them think about it?"

"We have to think long-term. Once this grass dies, it's gone forever, and so will their way of life. They should save it now. Build a new forest elsewhere."

Deka tried to take the tablet away, but Kylac hopped out of his reach. Deka grumbled and crawled under the entrance. "You're right, but hurry up!"

Kylac followed, still carrying his tablet. He and Deka ran through the spoke, hopping over so many bodies frozen in agony. Breathing in so much death and decay was already starting to make them edgy, and the mental noise these people gave off was just as unsettling.

The Relians sprinted down the paths, sniffing out the living through the dead. Kylac wanted to split up, but he knew the raptor would never agree. He hadn't left Kylac's side since Kronia.

They ran together from hut to hut, showing the symbols to everyone, reading their written replies. Everyone was scared and completely devastated by what they had witnessed. It was exactly like the first disaster, when the Archeons collapsed. On S'rin, everyone was linked together, so their minds had been pulled elsewhere, too. The shock had killed more than half of the population, leaving the survivors isolated and petrified.

At the fifth hut, the group wrote on a tablet and handed it to Kylac. The fox read it aloud.

"We felt where they went. They were pulled where there is no time."

"All the worlds he's destroyed, and he's no closer to controlling it!" Deka snarled. "It only took me a nine days to figure out how to control an offworld portal!"

"This is different," Kylac said.

Deka turned around and dashed out of the hut. Kylac lingered, puzzling over the message in his hand until Deka stuck his head under the entrance and huffed at him.

"Come on!"

Kylac dropped the tablet and followed.

The survivors were spread farther and farther apart. The disaster might have only pulled a thousand people out of this universe, but the shock was enough to kill at least three thousand more. What few remained were left huddled in small groups, dreading everything outside. The Relians explained it as best as they could to everyone they met, but writing took precious time.

At the eighth hut, Kylac entered first. As soon as Deka stepped in, he halted and stopped breathing. Kylac stared. He smelled it, too.

Deka approached the group, his nose leading him to the Tesha that smelled like Sonjaa. A Relian reptile's scent coming off a mammal was even weirder than coming from a Krone.

Meanwhile, Kylac held the tablet up, telling everyone who they were, what happened, where to find a portal off-world, and please think of the grass.

The people filed out of the hut, leaving Deka standing face to face with the bipedal rodent wearing Sonjaa's scent. She should have left with the others, but she stood still and stared at Deka. The reptile wanted to embrace her and hold her close and never let her go. He had done this on a couple worlds, forgetting he had to start over and get to know her each time, and it only startled her and made the process more difficult.

The Tesha broke away from Deka's gaze and searched the room, scrambling and jumping from wall to wall, but there did not seem to be any writing surfaces in here. Finally she approached Kylac and snatched his tablet. She pressed the surface clean and began clawing new symbols onto it. Deka waited patiently. He wished he could hurry up and tell her everything they had been through, but each version of Sonjaa had unique memories to sort out first.

The Tesha held the tablet up to Deka.

I know you.

Deka took the tablet and pressed a single symbol adorned with numerous subtle bends and dashes and accents on and around it. *I have met many who had your scent. You smell like my mate. She has been missing since the first disaster.*

She took it back and wrote. Deka stood beside her and read as she composed. *I know the Relian language. I am reading your throat movements. I remember meeting you on Rel, after you opened a portal under my feet. I remember you watching as I was laying eggs. I remember having a canine of my own. I have remembered these things all my life. Who are you?*

Deka leaned against her. She turned to him, reaching up to touch his snout. She ran her fingers over his scales.

She turned to the tablet and erased the top portion, and then wrote seven new symbols on it. She turned and held it up to him.

All my life I never felt right in this body. Since I was a child I have been traveling to meet other reptile species, trying to live like they did, even in the silence away from the other minds. I was searching for what was missing. It's you. Am I right?

Deka stepped in close and wrapped his neck around her as best as he could. The height difference made her uncomfortable, so Deka crouched. She embraced him, dropping the tablet, all six furred breasts hanging in the raptor's face.

As she felt his neck and shoulders, he bared his teeth and chanted quietly. "You won't die this time. Not this time."

Finally she released Deka, and he rose to full height. He couldn't help but smile at her body. He had always joked with fellow predators that mammalian breasts which remained filled even outside of pregnancy were a cruel twist of evolution—just a way of ensuring females were as immobile as possible so predators would have an easy time picking them off.

Now he thought Sonjaa's breasts looked beautiful and her body would be incomplete without them.

Deka raised his hand. She raised hers. They weaved fingers.

"I promise."

She leaned, picked up the tablet, and pressed a single symbol into it.

What?

"First, let's talk about you. Tell me everything."

He led her to the door. Kylac was already there, waiting for them. They ran together out the hut and back down the path.

4

"What does this mean?"

Stephen held the tablets up to Norh. The Krone glanced at them.

"The first reads, 'We've never seen anything like you before. When we have more time, we want to know you more.' The second is, 'We have been here for days, huddled in fear. You can run through the dark with no fear. We wish we could'."

"That seems—"

Norh spread his wings and took off. Stephen tried not to feel dejected. Norh never answered the same question twice, never explained himself "in other words," and sometimes that meant leaving questions unanswered. Stephen removed his backpack and shoved the tablets into it, making sure his clothes padded them.

Norh rose higher and higher, circling. The human wanted to ask how Norh knew where the people were, since he was too high up to smell them while they were completely enclosed inside those grass huts.

Stephen kept running. He was thirsty and hungry. On He'Kri, the last planet they had visited, he had eaten only seeds and water, and it had not been very filling, but he dared not slow down now. He promised to endure. To conquer. To succeed. Norh was watching, judging, and Stephen was determined to impress him.

As he ran down the path, Stephen got the feeling these huts weren't as spread out as they felt. If they only had a range of twenty feet, he guessed they were very close together, and there had once been many more of them, all linked together constantly. Now the people had been isolated in little pockets, afraid to venture into the void where there were no voices. Stephen felt privileged to be the one to link them together again.

He came to the next hut, crawled under, and ran straight for the group. Only five people stood here, three rabbit-things and two nonrabbit-things. They saw Stephen right away, and they walked toward him like a single person. It still looked unnerving. Stephen began dropping down to the ground to clear a space to write, but one of the rabbits held a blank square of clay out to him.

When he finished writing it, only the rabbit in the front looked at it. Half a second after he did, the other four began running. The rabbit who had read the symbols lingered for a moment, took the tablet from Stephen, made some marks on it, and handed it back to him before dashing out of the hut and catching up to the others.

As Stephen read the new symbols, he got the feeling he had seen them before. It reminded him of playing *Myst*: something seemed familiar, and a connection dangled right in front of his face, yet he couldn't quite find the connection.

Stephen shoved the tablets into his backpack, hoping he wasn't ruining them or his coat, crawled under the entrance, and dashed down the spoke. He stopped at the wheel. A few dozen people milled around like zombies, not looking at one another, or at the dozens of corpses littering the area. One saw him, and everyone turned and faced the human as he walked among them.

Several carried slabs of clay, and they handed these to him until he had a difficult time holding them all. He kept saying thank you, and he knew it was pointless, but it still felt rude to say nothing. The people ran back the way Stephen had come, climbing over the bodies of their fallen brothers and sisters back to the portal, leaving the human alone with an armful of tablets. He felt tremendously frustrated for not being able to read them. These people could be telling him the meaning of life, and all he could see was a bunch of stylized threes.

Norh landed in the roundabout. Stephen subconsciously stood fast against the wind he kicked up, still staring at the symbols.

"We've evacuated all the people from this region," Norh said. "I'll make another way further along the grassland."

Stephen stared at the digits. He was sure the resemblance to the number three was coincidence, but what if it wasn't? Forty-two was divisible by three, yielding fourteen, but what did that mean? Fourteen what?

He turned his face up to Norh. The Krone was looking down on him, wings unfolded from his back slightly. Stephen had had an even more difficult time getting used to Norh's way of smiling than Deka's.

He took out all the tablets he had collected and spread them out on the ground. Norh glanced at them.

"Any help?"

"The one in front of you reads, 'We felt the deaths of everyone. We wish you could feel them, too. They still cry'."

"Are you sure that's all they have to say to me? Seems a bit light for feeling thousands of people die."

Norh did not answer.

Stephen looked up at him. "Are you telling me everything? I'd really like to know what they're telling me, Norh. I feel weird surrounded by conversation I can't hear."

Norh met his gaze, wings still unfolded slightly. Stephen couldn't help but feel it was a condescending smile. Finally Norh folded his wings and turned away. Seconds later, a portal appeared in the circular clearing, and the Krone walked through. Stephen grumbled and took one last look at the tablets. He picked one at random, slipped it into his bag, and ran through the way. It closed as soon as he was through.

Norh was lying down, sniffing some of the grass. He reached out and grabbed a handful, yanking it free.

Stephen winced. "How does nothing hurt you?"

"I am a Krone."

"That's all you ever say. How does it happen? Why is nobody else immune to everything? What's the tradeoff?"

Norh was silent for a while, feeling the grass in his hand, scenting it, tasting but not eating it.

"Pryip," Norh said at last.

"What's a Pryip?"

"I want that to be our first planet. The people are immensely fascinating. A species of insect has a plant as its companion."

"A plant?"

"The plants are sentient," Norh continued. "The two trade brain chemicals to communicate. Insects sharing memories with plants. Try to imagine that. What would you do if you found out your only food source was intelligent?"

Stephen was still trying to figure out why Norh had brought it up. "I can't imagine."

"You can't imagine, and yet you wish to witness it anyway. You can't imagine life as a Krone either, and yet you still want to know. Even when you learn, you don't understand, but you still want to know."

"Right."

Norh tossed a blade of grass at Stephen. Stephen backed away, and it landed at his feet in a puff of dirt. Stephen looked up at Norh again.

"I smelled your fear."

"Yeah. I don't even have to cut myself on it to know it will hurt."

"Stephen, I understand what life is like for you. I understand what pain is, and how it makes you live. Can you not imagine a life without pain? Without fear?"

"Sounds like heaven."

"Yes, that place you mentioned when you told me about Jesus. You crave heaven, and yet you cannot imagine what it is, or what life there would be like. It is more an expression of what you desire out of life than what you believe to exist after death."

Stephen blinked a few times. "Norh, what are you talking about?"

"My species emerged during the earliest years of Kronia's development, when much of the planet was molten. Then frozen. Then molten again. My ancestors survived all of it. As a result, almost nothing hurts the Krone. It is only when we learned who our companion species was that we realized not everyone lives as we do. And then we learned nobody in the contacted universe lives as we do. All life is unique, but the Krone are special. You cannot even imagine it. Because you have no companion species, understanding something different from yourself is impossible, but I want to help you try."

"I'd like that."

Norh turned and faced Stephen, lowering his head to be at the human's eye level. "I can take you places Deka and Kylac would not. I want to take you to so many planets that you can experience life as every species in the universe, out of hope that it would build a new perspective within you. But you are not ready. You still fear me. When we go somewhere on our own, you cannot be afraid of me. You cannot be puzzled by anything I say or do. I will know when you are ready for it."

"Of course I'm afraid of you. You can stop lava flows and dig through dry ice and touch sharp grass and not think twice about it. Look at me? What the hell can I do? I thought I felt inadequate compared to Kylac and Deka. Now I meet you, and I feel even more useless. Compared to you, I can't do anything."

Norh's wings unfolded all the way to the ground. "I am no god. When you are able to think of me as just another person in the contacted universe, you will be ready."

A portal opened where the previous one had closed. It led back to the hub. The other portal leading to the bird world was still visible, and a steady stream of rabbit-things and nonrabbit-things were walking from Kylac and Deka's portal against the wind to safety.

Norh extended a claw, scratched three new symbols into the ground in front of Stephen. "These are the new symbols." He stood, flapping his wings, and lifted off. Stephen took out a tablet and copied them. He looked up and waited for Norh to start circling.

As long as Norh had been with him, Stephen still did not know the Krone any better than the day they had met, and yet the Krone claimed to know Stephen better than he knew himself. He kept insisting he was nothing to envy, and Stephen had no idea what he was talking about. Norh had everything to envy. He was not surprised people came from all over the contacted universe to meet the Krone. They were the embodiment of freedom and power, with or without portals.

The Krone rose high in the sky and circled. Stephen took off running, hopping over many green bodies littering the ground. The stench was difficult to ignore. He realized there were fewer survivors here, and more bodies clogging the paths through the grass. Stephen figured normally these people would have been out and about, everyone connected to everyone else across the whole population of over six thousand. A single mind.

Stephen wondered why they simply did not wander around until they found someone, but he resisted the urge to ask Norh this question. He tried to imagine what it would be like for them, constantly used to the thoughts of six thousand people running through their minds all the time. To them, being without that must be like when Stephen had been on Eiae without light.

Likening it to walking around in the dark helped a lot. Stephen could imagine that. He could imagine why they huddled together in tiny groups in the huts, terrified to leave the safety of each other's thoughts and venture through the void.

Stephen grabbed several bodies and pulled them away from the entrance to this hut. These people were heavy, but gravity was slightly weaker than Earth's thanks to the planet's missing mass, so he could lift more than he was used to.

With the entrance uncovered, Stephen crawled inside. The people had already gathered in front of it, waiting for him, all six of them staring blankly. Stephen held the tablet up. As one, they ran around the hut and picked up one nonrabbit-thing and carried it out with them. Two picked up pieces of clay, etched some things onto them, and passed them to Stephen. They then joined their companions carrying the unconscious nonrabbit-thing out of the hut.

Stephen looked at the first tablet.

3 3 3 3 3 3 3 3 3 3 3

3 3

3 3 3 3

The second read:

3

3 3 3 3 3 3 3 3 3 3 3 3

The digits were covered in seemingly random curves and marks and dots and dashes. Stephen was sure he knew this. He'd been staring at symbols like these for hours, and the feeling he got as he read this tablet was: *I saw Norh. He is laughing at you.*

Once he saw it, he knew it. He read the symbols, recognizing which feeling radiated from each. Stephen gasped as the wind shook the hut so hard he felt like an ant in a hurricane. He turned to the second tablet, and now each symbol leaped out and spoke to him.

Deka and Kylac laughed at you when you first got here.

He turned back to the first tablet and read the symbols again. This time he noticed some things he had missed, some contractions that had not been obvious to him just a moment before.

You see all of them as gods. They see you as an ant. Norh doesn't need you. He can help the survivors himself. He's keeping you busy.

Stephen stared at the tablets for some time, reading them again and again. The same messages popped out at him. He stared off into space, wondering about the telepathy. Was it contagious? Were the Archeons wrong about these people not being able to speak to others? He was an unknown species, so it was possible they didn't know what effect it would have on him.

Suddenly the roof of the hut tore off and flew away. Norh landed on the dirt floor in front of Stephen.

"Why do you delay?"

Stephen was still holding the tablets. He slowly turned his face up to Norh's. "You don't need me, do you?"

The Krone regarded him for a moment. "I can reach the survivors myself and warn them, but I will scare them. You are less likely to."

Stephen stared at him. "So they were right. How did they know?"

"How did who know?"

Stephen held up the tablets up. "I can read it. I don't know how, but I can read it. You're patronizing me, aren't you?"

Norh glanced at the tablets, then met Stephen's eyes. "Leave the tablets. I have them memorized."

"I already know what they say. Norh, they're teaching me their language mentally."

"They can't."

Stephen wanted to insist, but he thought better of it. He walked away silently, crawled under the entrance, and jogged back to the wheel. He waited there until Norh started circling, then ran down another spoke. The going was slow, and Stephen had plenty of time to ponder the tablets as he climbed over so many bodies. So much death. So much fear.

5

"I was just born when I felt it for the first time."

The Tesha's words were soundless. She had no vocal cords, so she had no voice, but she had learned the Relian language, as well as many others, by imitating mouth and throat movements. This was what Deka and Kylac were watching: her throat, tongue, and mouth moving correctly to form each word.

"Everyone noticed I was surprisingly comfortable away from the minds of the others. I yearned to feel the silence other species had to deal with. Even when we would travel offworld, I would often separate myself from everyone and try to live like they did. Then one day I met my first reptile species..."

It had been like learning Relian all over again, but now there would be no more writing, no more long delays between question and response. Communication was now as seamless as if she were speaking.

They were sitting in one of the circular clearings, among the twisted bodies of the dead. The stampede of Teshi and Tesha had trampled the bodies so much they looked as if they could never have been alive.

This area was empty of survivors. Kylac was calculating a new way to another place on the subcontinent. The

Tesha filled the silence as both Deka and Kylac began hearing Sonjaa's voice coming from her mouth.

"...the Gelleens."

Deka smiled. Kylac's tail thumped the dirt. The Tesha rubbed her hands together, scraping imaginary claws up and down. She did not seem to be imitating it. It was as if she really laughed this way.

"I stayed with them for a long time, learning their language, learning how to listen with my eyes and speak with my mouth. The skill came easily to me. So easily I became the speaker and listener for many visits offworld, and for many people who visited us."

"For how long?" Deka said.

"About twenty years."

Deka glanced at Kylac, each reading the other's expression and scent. Both were thinking the same thing: they had been to S'rin only five years ago. They had been to Kronia many times, and all these other worlds, so why had Deka never met or heard of any of these unusual people before?

The Tesha's name had no translation. The people of S'rin had no names, only unique mental vibrations, so Deka did not know what to call her. Everything in him wanted to call her Sonjaa, and in his mind he had already started to refer to her that way, but he hesitated to speak it aloud.

"I sought out the ways to worlds with reptiles. I felt more comfortable with them than with my own people. I even coupled with a few different reptile species. I never understood it, and nobody on S'rin understood it either, but it just felt right. Then I smelled you. I saw you. It's you..."

"You recognize me?" Deka said.

"I don't recognize you," she corrected. "I *remember* you. Sometimes my memories of Rel are more real than the ones of S'rin. They were a puzzle to all of us. Nobody knew where the memories came from, or what they meant, but when you ran into that hut..."

She crouched closer to Deka and reached out to touch his arm.

"The way your scales feel in my hand... It's the way skin should feel. Not furred and smooth, but—" She stroked Deka's arm. "Like this. I remember this. Now I finally know why it always felt right, and me being in this body felt wrong."

She looked into Deka's eyes as she felt his arm.

Deka swallowed, hesitated a few times, and then finally spoke. "What is your name?"

"Sonjaa."

Deka heard her voice. It had been rising in clarity as she spoke, and now it sounded perfectly clear. Deka reached out to her. She was covered in fur, her snout was wrong, her hands had pads, her fur was smooth and clean, and she had breasts. She was everything Deka was not attracted to, and yet everything about her felt right.

Deka leaned into her, wrapping his neck around hers. She embraced him back in the same way. Differences in biology limited how she could do it, but she did not seem to be imitating Deka's movements. Everything she did felt natural and normal for her.

He pulled away to face her. "What is your last memory of Rel?"

Now she hesitated. Eventually she found the words again. "I remember seeing you. I remember Friend and Rive and Kylac. There were so many people around me. Around us. Relians. It was not on the hub. I was in the air. Something was turning me around. Pulling me in every direction at once. I screamed for help, and then the memory stops."

"Do you remember why you were there? Why we were there?"

"I'm not sure. Everything just ends there."

A way opened. Kylac stood and walked through. Deka and the Tesha lingered on this side for a moment longer. Deka breathed her air. She breathed his. It was as if no time had passed.

She leaned forward, opening her mouth, and nipped the tip of his nose. Her teeth were smaller and sharper than a raptor's, but the effect was the same. Deka leaned closer to her and nipped the side of her neck.

Her scent changed. Instead of just Sonjaa, now Deka smelled arousal.

"It's a Relian reaction," she said, "I know—I've always been this way. It's why everyone thought of me as a reptile trapped in a mammal's body. It was a joke among the community all my life. Now I know it wasn't a joke. I remember biting you there a lot."

"Do you remember why?"

"It's where I bit you after I fell through the portal you opened under me the day we met."

She rubbed her hands together, clicking the nails. Deka rubbed his claws.

"And there I was calming you down while bleeding from the snout."

"I was happy with just a fox. Then you talked your way into my life and made everything complicated."

Deka smiled with his hands as they walked through the portal together. Kylac had made it Krone-sized out of habit, so they both fit through easily. They now stood in the middle of a scorched landscape. The grass had been charred to cinders here with nothing left standing, and the bodies of hundreds of Teshi and Tesha had been burned so badly they could not tell the two apart.

6

Stephen wrote six symbols on the tablet and showed it to the survivors in the hut, asking if they knew who he was.

One of the rabbit-things grabbed the tablet and clawed it.

We know.

Stephen etched six more variations of the number three on the tablet and showed it to them.

"What is Norh planning to do to me when we start traveling on our own?" he said aloud.

They took the tablet, made some markings, and handed it back. The group of eight dashed past Stephen and under the entrance, leaving the human alone with the decaying bodies. He looked at the symbols.

Anything that amuses him.

Stephen shoved it into his backpack and crawled out of the hut. He scurried back to the roundabout and wandered in thought as he waited for Norh to rise again.

A sound came through the grass. Stephen faced it, bracing himself to run. The grass rustled, feet stomped, and a nonrabbit-thing tumbled out and rolled to a stop on the dirt just a few paces from the human. It was bloody from head to toe, covered in long gashes where the grass had flayed it open. Stephen gasped and ran toward it. He wasn't sure if touching it would be a good idea, so he leaned over it. It looked at him, and Stephen heard it screaming in his head.

He dropped his bag, took out the tablet that told where the exit portal was, but he thought twice. Instead he took out one of the others he had saved and worked the surface to a blank slate again.

"Who are you?" he said as he carved the symbols. "What happened to you?"

He showed it to the green-furred insect. It painfully took the tablet from Stephen and cut a few new symbols underneath Stephen's.

He has left you.

Stephen couldn't catch his breath. He carved six new symbols and held them up to the nonrabbit-thing.

"What's he doing?"

It carved nine, sloppy threes in reply. They were hard to make out, but once Stephen recognized them, the meaning took him by the throat.

He's setting our huts on fire. He's laughing at us as we run. I survived by going through the grass.

Stephen stood and scanned the sky. Sure enough he saw a plume of smoke in the distance, and some flickers of flames. He noticed the air had fogged with smoke. He still couldn't catch his breath.

Stephen picked up the nonrabbit-thing and brought it to its feet. Now he was sure this thing was the result of an insect crossed with some sort of furry mammal, as its skin did not yield to the touch.

His long underwear shirt was bloody now, but he didn't care. He walked forward, hoping the thing draped over him would take the hint. It began walking with him across the wheel and back down the spoke that led to the portal.

Moments later, dozens of rabbit-things and nonrabbit-things flooded in from the spokes and overtook them. Their silence disturbed Stephen; he was used to screams and shouting when in a panicked group.

A female rabbit-thing slipped under the other arm and helped Stephen walk the injured nonrabbit-thing. They moved so fast they practically carried it between them for several strides.

Stephen was in the middle of a mass of people fleeing the approaching fire. They had to climb over many bodies of the fallen, and it slowed everyone down, especially Stephen.

Someone stuck a tablet in Stephen's face.

3 3 3

He only needed to glance at the symbols now before he recognized the meaning: *The portal is gone! Run!*

Stephen tried to run faster, but the group wouldn't speed up. Stephen shouted for them to hurry, but for all his shouting they never reacted to it.

Norh dove from overhead, grabbing grass in both hands and both feet, tearing an entire city-block's worth of plant life away. He banked and made another pass.

Stephen didn't need a tablet to figure it out—Norh was hunting them! He had started a fire to herd them into a place with no escape, and now he was going to pick them off, first the wounded, then everyone else. Stephen clutched the victim tighter. He would not let Norh take this one. He would not let him win!

Stephen pushed on over the fallen bodies. Meanwhile Norh flew behind them, swooping back and forth, tearing the grass away. Then he made one final pass and landed between them and the fire, tearing still more grass out.

They ran through several wheels, down multiple spokes. The group knew where to go, so Stephen ran with it. He began to hear screaming, but nobody's mouths were open. He heard shouting all around him, and it carried him faster and faster as the wind pushed the fire to them.

A pile of bodies down one of the paths held the group up, and Stephen used the moment to look back.

The fire was still gaining on them, and there Norh stood, inside the flames. He was ripping up flaming clumps of grass up and casting them aside. Norh saw Stephen looking at him.

"Keep running!" he shouted from the fire.

"He wants us to run," Stephen muttered. "He likes his prey to run. He *wants* us to run! We should stay put!"

The group couldn't hear him, and they pushed on around Stephen. The female rabbit-thing helping him carry the grass victim urged Stephen on, but Stephen pulled backwards.

"He's trying to make us run! It's what he wants! Stop! Stop running!"

The rabbit-thing kept pulling and pulling. Now even the victim was trying to urge Stephen forward. The people pushed him, and Stephen wished he had some clay or a blank piece of ground to write it down. He knew the symbols he wanted to make—he knew what he wanted to say— if only he could talk to them mentally.

Stephen tried it. He tried thinking loudly. Four symbols—he screamed them in his head. *Do not run! He wants us to run! Do not give him what he wants!*

A brief moment of hesitation. Then someone knocked Stephen away, took his place under the wounded person, and carried it forward. Everyone climbed over Stephen as if he were a dead body. Their feet hurt him, especially the nonrabbit-things, as they had sharp claws instead of nails covering their fingertips. When they had passed, Stephen finally rose to his feet.

He stood alone, facing the approaching fire. Norh was still between it and them, tearing the grass out, prolonging the fire, keeping it away from them so they had more room to run. Stephen refused to play the game. He stood resolutely.

Norh turned and faced Stephen from the flames again. This time Stephen met his eyes, and he felt no fear.

It was too far away for Norh to hear him, but he could read. Stephen bent down and drew five symbols on the ground.

"I know what you're doing. I won't play."

He rose. Norh still glared at him, holding the fire back, daring Stephen to run.

Stephen took off his backpack and held it out. He would run all right, but not where Norh was herding him. Stephen bolted headlong into the grass.

"Stephen!"

The human felt proud of himself for finally breaking Norh's omniscient demeanor.

7

Deka and Sonjaa walked together across the burned land. What was left of the grass crunched under their feet. They had left the path long ago, and now they wandered through the emptiness, searching for anything that remained.

Kylac was on all fours far ahead of them. Deka didn't know what the fox was scenting for, but he trusted the fox to handle himself out here.

Fire was a constant threat to these people, and the usual response to it was to cut down the grass between them and the flames and contain it before it engulfed the entire continent. It worked, but this time there had been no warning, and not enough people. Everyone had died alone and scared.

Deka felt a padded hand on his flank and turned to meet Sonjaa's eyes.

Her mouth moved. "What started the fire?"

"Magma," Deka said. "The disaster pulled the planet apart, scattered magma everywhere. We were too late to help these people."

"Nobody thought of that," she said. "The groups I was with. That's what I did after the disaster struck us. I was the only one brave enough to move from hut to hut, spreading news and the comforting thoughts of others."

"Why didn't you bring them together?" Deka said.

"I tried to tell them where the others were and we should be together, but they were too scared to move. So I ran between huts, comforting everyone I could."

"How long ago was the disaster?"

"Just days ago."

"Perfect. We could catch up to Friend with the next portal."

"Friend... I remember that name. And Rive. I also remember someone named Taris. Friend's mate?"

Deka stopped walking. "You remember Taris?"

"Yes. We used to watch as Friend hunted for her. Rive didn't care much for hunting. He preferred to sit and think about big ideas." She rubbed her nails together. "Never understood how he kept Friend from the old ways when he let his fox kill."

Deka took her hand, rubbed his claws against her fingers, laughing with her—it had been so long since he'd laughed with another raptor. He never realized how starved he had been for the company of another reptile from Rel.

"I asked Rive about that many times. Rive, you're not much of a raptor if you let your fox hunt all the time."

"He never bit you back," she said, silently, laughing with Deka. "He knew it was true."

"So did Friend. He always felt weird as the only fox on the hunting grounds."

"But he could run with the predators," she said. "Do you think that's why he's in such trouble now?"

"Rive was never a very strong raptor. Having a fox who didn't need much taming seemed to be the perfect match. Now he can't stop his own fox."

Kylac was running toward them.

"Did you find something?" Deka asked.

Kylac slid to a stop and rose up on two legs. "It's all burned, but I can smell where the fire is now. We might be able to reach them before it hits."

Deka turned, grabbed the Tesha by the shoulders, and lowered his head to be at her eye level. "Sonjaa. I want you to wait here. Don't move from this spot."

"Alone? With the dead?"

"There's no crowd here, but there will be one in front of the fire. Every time I've met you, you die in a group. I've met you so many times only to have you ripped away—but not this time!"

"If there's a fire, I can help! Writing to the people will take too long. I can reach them directly."

"If you go with me, you will die!"

"Deka, have a little more faith in me than that. I can hunt with any reptile in the contacted universe."

Deka glanced at Kylac. Kylac wagged his tail and turned away, concentrating harder to speed up the calculations.

"I believe you," said Deka. "But I have been trying to save your life since the first disaster. Every time I meet you, something horrible happens. You will die if you come with us."

"I may be a reptile in a Tesha body, but these are still my people. If I can help them, I will."

"Sonjaa... A crowd will separate us, and I won't be able to reach you. I don't understand what's happening, but I won't let it happen again. Please stay here."

She stood still. Her body language was so atypical of a Tesha Deka forgot she was covered in fur. He knew exactly what she was going to do.

The portal opened in front of Kylac, Krone-sized again. Sonjaa bolted past Deka, breezed by Kylac, and straight into the sphere. Crying out, Deka turned and ran after her.

8

Stephen's backpack protected his hands and chest from the grass for about thirty strides, and then it fell apart. His winter coat, long underwear pants, binoculars, and a few tablets spilled out. He grabbed what he could and pushed onward through the underbrush. It was just like trudging through the wooded areas between the housing developments back home, except the grass sliced through his skin here.

The grass had been whittling his pants down to the threads, and now it started to kiss his legs. It snagged the long underwear shirt, pulling threads from it, leaving them unwinding for many strides before they finally snapped.

His shoes were falling away. He began to feel bare ground beneath his feet, and any blades of grass he stepped on cut his soles, heels, and arches.

The smoke from the fire blew overhead. The heat from the flames embraced him, and it felt so good. He wanted to slip on his coat and long underwear pants, but there was no room to move in here. The grass was so thick it was a wonder he had made it this far in.

He held his coat, using it to pad his hand as he pushed the grass away. The blades sliced through in just a few steps. Stephen bled from his arms, legs, and face. He imagined he was just as torn up as that nonrabbit-thing. That survivor was Stephen's hero. He had survived by refusing to be herded. That's how thinking creatures survived. Not playing by the rules because the rules were made by the predators for their own benefit. Stephen had won. No matter what happened, Norh would lose.

Many more paces into the grass, Stephen began to wonder if it ever ended. He thought that about his many walks in the north country. Even the wooded buffers between housing developments seemed to go on forever, though civilization was just a few feet away.

A Krone flew overhead. Stephen crouched low and held still. He heard the fire coming closer. Stephen realized Norh expected that, too. Stephen was supposed to run from the fire. Of course! How stupid of the human to fall right into that trap! Stephen turned toward the smoke and ran into it.

"Stephen!" Norh shouted from the sky.

Stephen kept running as fast as he could. The grass sliced his long underwear shirt to ribbons and then started splitting his skin open. The flames were in sight. Freedom was in sight! Freedom from the game!

Something landed on him and flattened him to his stomach, knocking the wind out of his lungs. Stephen struggled and screamed and reached out for the flames. The grass in front of his eyes turned black and shriveled. Orange flames broke through them and washed Stephen's face in heat.

Claws surrounded him, and so did the fire. Norh squeezed his entire body and lifted him up. The ground fell away with a sickening lurch, but suddenly Stephen didn't have the energy to fight back.

Stephen now saw S'rin from a Krone's point of view. The grass never ended. It covered everything. Paths cut through it, round circles, more paths ending in huts. Stephen now knew it wasn't as big as it felt while he was inside it. The fire burned through everything. In the distance, Stephen saw more fires, all coming from fallen columns of the broken planet.

Norh rose higher and higher, and then he hovered in place at what must have been three hundred feet up, out of the smoke, out of the fire. The air was clean up here. Stephen's mind cleared. Norh lowered his head to talk to Stephen as he flapped in place.

"What did you think you were writing?"

Stephen was out of breath. "I know what you're doing. I won't play."

"The symbols you wrote read 'Displays resulting in several regarding time jointly. Water generally split up gets older into underhanded educated to certainly'."

Stephen hung there in Norh's claws. Suddenly he didn't remember why he wanted to get away from the Krone.

"What did you think the tablets said?"

Stephen swallowed a few times. Everything on him hurt. Blood was dripping off him far down to the fire below.

"They were telling me you were hunting them. They said you didn't need me. They told me you started the fire... And the portal was gone. You were herding us. Hunting us for fun."

"Where did all that come from?"

Stephen thought about it for a moment. His head was so much clearer up here.

When Stephen didn't answer, Norh answered for him. "You thought you could read their language."

"I saw the symbols. They started making sense."

Norh soared down. Stephen closed his eyes, shielded his face, and the stinging wind bit into the slices across his body. He wanted Norh to slow down, and just when he thought he couldn't take it anymore, Norh landed and set Stephen down. They were in front of the portal to the hub. Stephen just realized he was wearing only what was left of his long underwear shirt and pants.

Stephen scanned the area. The fire was all around them, but it hadn't touched this grass. He remembered from his time in the air that this part of the path had been free of grass. Now it made sense. Norh had ripped it all up, keeping the fire from cutting everyone off from the portal.

Norh lay facing the human. Stephen sat on the dirt, wincing in pain every time he breathed.

"Remember Eiae?" Norh said.

"Yeah..."

"Remember what Deka and Kylac told you? Electromagnetic radiation is all around you, but this time it's generated by the Teshi and the Tesha. It does affect you. To your mind, it is noise."

"Noise? Oh no. I heard them screaming in my head. I could read what they were writing. I thought I could..."

Stephen raised his eyes to meet Norh's. The Krone's face was so rigid, so Stephen observed the position of his wings. They were down. Stephen couldn't tell if he was tired, depressed, or disappointed.

"The ones in the hut were pieces of the same message," Norh said. "What did you think that was?"

He still remembered it word for word. "I saw Norh. He is laughing at you. You see all of them as gods. They see you as an ant. Norh doesn't need you. The grass doesn't harm him. He can help the survivors himself. He's keeping you busy."

"Where did those words come from?" Norh asked again.

Stephen blinked a few times. He did not break eye contact with the Krone. "From me."

Norh raised his head. "Your brain was aware of the mental noise, trying to make sense of it. It is how the brains of most species work. The mind tries to attach order to chaos. A familiar, orderly world is a predictable world, and a predictable world is a safe world. Your mind reaches for these connections whether you know it or not. Because you're an uncontacted species, it is uncontrolled, and your mind frequently finds connections that are not there."

"I knew you didn't need me to warn the people. I thought you started the fire. I thought you were herding us to our death."

"You saw the fire but did not comprehend it right away. Your mind integrated it into the orderly world it was trying to create out of the chaos it was picking up. It incorporated every fear you had."

"I was going to run right into the fire just to escape you."

"It usually does not have this effect on anyone, but there is a lot of fear among the Teshi and Tesha. Normally it only makes offworlders feel unsettled. This is why you are not ready to be alone with me."

Stephen shivered and averted his eyes. He felt like a teenage girl having her diary read at the prom. "I can't do anything out here. All of you can do everything. I feel like an ant... and you guys watch me scurry around trying to avoid the magnifying glass."

Norh's wings lowered to the ground. "I know you think of me that way. It's in your scent and your body language. Were you not aware of it yourself?"

"I didn't think it was... this bad."

"I already knew. Deka and Kylac know. The only one who must accept it is you. Once you do, you can overcome it, and then you will be ready to understand other species."

Norh rose and walked past Stephen. "I will tell you the rest of their messages and clean out those wounds on He'Kri. Their written words contain many layers of meaning. Interpreting them requires practice, and learning how to think in an entirely new way."

He passed through the portal and paused on the other side, waiting for Stephen. The human slowly climbed to his feet, feeling worse than an ant for letting Norh down. He stepped through the sphere and followed close behind Norh so he could stand up in the wind.

The portal closed. The fire raged on. It would burn for days before the air bled off into space.

9

The fire on the horizon devoured the grass, driven by the fast, constant wind. Deka and Kylac stopped at another hut. Sonjaa crawled under the entrance and was inside for only a moment before the people began pouring out and down the path.

"Amazing the fears you can overcome when someone tells you where the exit is," Kylac said.

Deka's agitation had not waned since he set foot on this region. He remained silent until Sonjaa crawled back out, and they ran with the people to the clearing. The people ran down the spoke and toward the sphere, and the Relians paused to catch their breath.

It was the tenth hut they had found. They scented for another, but as the fire came closer, scents became harder and harder to make out. Kylac managed to pick out some of the living scents among the bodies and the fire and led them down a spoke. It ended in another clearing, and Kylac followed his nose down yet another spoke, then to another, and then to another. The fire was very close now.

"Is this the last?" Deka said.

"As far as I can scent, yes. If there are others, we'll never reach them in time."

"Good." Deka glanced at Sonjaa. He rubbed his claws as he ran. "For someone with breasts as large as yours, you run very fast."

She held her hands together, rubbing her imaginary claws. "I'm still not used to them. They never felt right to me. I asked someone to remove them once, but she warned me it would create more problems later in life."

Deka stared at them. "They look good on you."

She leaned over and nipped him on the snout. Deka feigned injury.

The path ended at a hut. Kylac bent low and scented the air coming out of it as Sonjaa crawled underneath it.

"Deka..." she called through the glued grass.

The raptor crouched and peeked under. His mouth hung.

Over four hundred people had crammed themselves inside, and Sonjaa was lost among them. Deka scented the air, smelled the general direction where she was, and shouted.

"Sonjaa! Hurry!"

Then he remembered she could not hear him. He ducked under the entrance and stood to his full height. Several breaths passed, and the sound of the fire grew much too close. Deka's heart raced as he sprinted around the group huddled in the center. They squeezed together so tight Deka wondered how they were breathing.

"Sonjaa! What's wrong? Why aren't these people moving?"

Desperately he scanned the hut, but he had lost her.

Kylac smelled the fire. The smoke was getting much too thick, and the flames were almost on top of them. He crawled under the entrance and ran around the cluster of tightly packed rodents and insects.

"Sonjaa!" Deka called. "Why are they standing still?"

Kylac saw her first. Somehow she was in the middle of the crowd. Her lips were moving, and she was screaming, expecting Deka to hear her.

"She says they just got together they're not going to split up now! They're too scared. Sonjaa! The fire is here!"

Deka found her. She was facing away, but somehow Deka could hear a voice. "I'm reasoning with them," she was saying.

Deka leaped into them, trying to push people out of the way, but they were as unyielding as a leaf on Kattaaka.

"We're out of time!" Deka shouted. The people were closing in, packing themselves in tighter.

The back wall of the hut turned black. The far side of the roof dissolved into smoke. Orange and red flames crawled in and devoured the grass.

Kylac pushed and shoved and tried to pull people down, but they had become stone fixtures. The billowing smoke burned Kylac's eyes and overwhelmed his sense of smell. When he lost that, he backed away and ran for Deka.

Deka raised his foot and slashed the Tesha female in his way, but his claw did not seem to do any damage. Deka slashed again and again, using both of his hands and one foot, but he smelled no blood, felt no penetration.

Kylac grabbed his arm and pulled him away. Deka was standing on one foot and fell into Kylac. The fox caught him, stood him back up, and yanked his arm toward the exit. Through the smoke, Deka saw Sonjaa, immobile within the panicked mass, unreachable. Her mouth was just barely visible, and it was still moving.

"...won't leave! They won't leave! It doesn't make sense! They're just afraid! They won't leave! I can't get out! Deka, help me!"

"Climb!" Deka screamed. "Climb out!"

She didn't see him.

The smoke was so thick it hurt to breathe. Deka turned in the direction Kylac pulled and ran with the fox around the mob, ducking under the entrance and rolling out.

The two coughed and gagged as smoke and flame surrounded them. Now Deka grabbed Kylac's hand and led him through the spoke, flames licking the grass on both sides. They jumped over dead bodies, nearly blinded by the smoke both in the eyes and the nose. Only Deka's unified mind kept him orientated and running toward where he knew the portal was. Pieces of burning grass fell on the path before them, and Deka tore through them, ignoring the pain. Scales on his back burned, but he ran, teeth clenched, hand in Kylac's, toward the portal.

10

The Teshi and Tesha refugees had gathered under the trees of He'Kri. Some of them held up seeds as they wrote on pieces of clay. They passed the seeds back and forth between them and the birds of He'Kri. The symbol for grass was used frequently.

The Teshi and Tesha herded within their large group many species of animal they had rescued from their dying world. They dared not set them free, not yet. They still didn't know if they would be able to survive here, and they had not yet decided if this would be their new home, or if they could find a more suitable place to live.

Norh licked Stephen from head to toe. The human was covered in deep gashes where the grass had cut him open. Dirt, sweat, and shards of grass were also buried in his skin. Norh was working them free right now as Stephen screamed, bracing himself on one of Norh's claws.

His only possession was his boxers. They had somehow survived his trip through the grassland. He was worried how he would stay warm on cold worlds, but for now, he was too focused on the pain to be concerned about that. Several times Stephen glanced back and noticed Norh's wings outstretched. He wondered if tasting good to a Krone was a good thing or a bad thing.

Kylac sat on a wall behind Stephen and Norh, watching Deka in the treeless field enclosed by large stones too high for He'Kri's predators to jump out of. The flesh-eating birds gathered and confined their predators inside these pens. Not only did it allow them to eat without having to hunt, it meant the predators would not be able to climb the trees and harm the seed-eating birds.

Deka was in the pen, sprinting after one of the feline-like things. They were fast, but not nearly as agile as Deka. He caught it, pinned it, shoved his claws into it, screaming.

The animal was long dead, but Deka reached in with both hands and pulled it apart. He shoved both feet into it, scattering its guts and flesh and bones every which way, showering both himself and the stones in blood. The other cats huddled around the perimeter, shivering.

When there was nothing left to tear apart, Deka lunged for the next closest cat. It scampered away. Deka chased it, then another, then another, then another, then finally pounced on one and tore it to shreds. Deka washed his face and body in its blood. He reared up and screamed.

Stephen echoed the sound as Norh worked his tongue over another wound and pulled out a long fragment of grass. Kylac looked back at them. Stephen was holding onto Norh's hand as the Krone's tongue worked each cut open and cleaned it out.

The fox turned to face Deka again. The raptor was still screaming, now picking up bones and crushing them in his jaws.

Kylac's ears remained folded against his skull. Watching Deka vent his rage and hearing Stephen vent his pain made Kylac wish he could vent, too. He scanned the birds, the Teshi, and the Tesha, sniffing out someone who might be interested in a Relian canine. Someone who wasn't too busy discussing grass seeds.

Hypsil

As soon as Stephen came through and felt the chill in the air, he wished he had more than just boxers to wear. He stood close to Norh, letting the Krone's scent kick up his metabolism.

Deka and Kylac stepped through next. The daytime star still hung high in the sky, but the light was too dim, the energy it gave off too cold.

This should have been the hub, but as usual nothing was here. They saw something strange a few dozen paces in the distance, and as one they began walking toward it. The wind was cold and strong, so Stephen walked in front of Norh, and the Krone breathed on him. This method had been how he endured the cold since losing his clothes.

They stopped at the ledge and peered over. Instead of the science model cutaway of the planet, they stared down into blackness. Not the blackness of space, but the pure emptiness of a place where time ceased.

Norh reached forward and grabbed Stephen around the waist as he spread his wings and took to the air. Deka and Kylac watched them ascend. Stephen didn't even scream this time. Deka stood with his hands together, and Kylac's tail swished.

The raptor and the fox observed the weak star and the black, timeless landscape. Deka growled inwardly, breathing through clenched teeth. He sat and snarled at the ground. Kylac huddled against his raptor.

"I have something new I can try," Deka said. "If she's here this time."

"These last few thought they *were* Sonjaa," Kylac said. "They knew everything about her. About us."

Deka snarled louder. "And she keeps dying! I know it's coming! I'm ready for it, but I can't get through the crowd! What is stopping me? Why can't I help her?"

"Calm down and think, Deka. Maybe preventing her from dying is impossible."

"I lost my entire species! I won't lose her, too!"

"You've tried fourteen times. On the last world, you just thrashed, and it still didn't help. You should step back and save your energy. It might help us figure this out, and then we'll know what to do."

"She's my mate! I won't stop trying. Something has to work."

"Think about this rationally. All the incarnations of her agree on a few things. Somehow she remembered seeing you as she died. Rive and Friend were both there at the time. I don't understand how, but I think we're getting closer."

"How do you figure?"

"We're meeting her more frequently, and each person on each world remembers more and more. We could be approaching the epicenter of whatever happened to her. I think we'll know it because we'll meet her. Not just any her, but the one that's a Relian."

"Maybe then I can finally pull her out before she dies!" Deka snarled again. He reached out and clawed the dirt. "We'd catch up to him if we didn't have to keep cleaning up his messes! As soon as she's safe, I will hunt that canine myself and kill him, and then I'll fucking slit Rive's throat for letting this happen over and over!"

Kylac ran his hand up and down Deka's neck. "Don't wear yourself out trying, and don't do anything we'll regret. We still don't understand what's happening. We still don't know what to do."

"I've seen enough." Deka sat still and breathed for a moment.

The star in the sky flickered. Matter flung off of it. The cold wind continued rushing by. Moments later, a Krone landed beside them on three legs, setting Stephen down gently with the fourth. He did not turn, but kept his eyes on the horizon.

"It's another negative lens," Norh said. "It's equal to the missing third of the planet."

Kylac's ears swiveled back and forth. Deka slowly turned his head and stared at the Krone. Stephen stood and gawked at the horizon.

"And there is more," the Krone continued.

A Krone-sized portal opened behind them. Norh turned around and walked through. Stephen followed, chasing the heat more than the adventure. Deka and Kylac turned around and gazed at the view projected around the outside of the sphere.

Skeletons.

They entered.

The valley was a mile deep by Stephen's measuring scale, and they stood halfway up it, on top of a bed of dry skeletons. The portal Norh opened had set them down in the middle of it.

The wind was weaker here, and Stephen ventured away. His feet had toughened up since he lost his shoes, and now he walked on rocks and bones without wincing. He went twenty paces into the silent valley, bent down, picked up a skull and held it in his hand. He turned and watched his three companions as they took their first steps out into the valley of skeletons. Kylac was on all fours, scenting. Deka bent low, also taking in scents. Norh remained upright and regal, turning his head back and forth, surveying the land.

When Kylac had taken in all the scents he could, he rose. "They've been dead for years."

"Years?" Stephen said. "How?"

"I don't know."

"We should go to each region," Deka said. "There may be survivors somewhere."

"Who were these people?" Stephen asked.

"The Hypsil are quadruped herbivores," said the Krone. "They lack claws, sharp teeth, or armor. All they had were the Hrha. Carnivores. They herded the Hypsil into large groups to protect them from other predators so they themselves could hunt them. The arrangement brought them both to sentience."

"Really? How?"

"They actually reached it together," Deka continued. "I know what you're thinking. The society would have developed so the carnivores oppressed the herbivores, kept them stupid, used them for food."

"Well, yeah. If your society grows up with one race dominating the other, you start thinking that's the way it's supposed to be."

"Believe it or not," Kylac said, "it didn't happen that way. Once the herbivores learned the people who ate them were intelligent, they started looking up to them. And once the carnivores learned their prey was intelligent, they stopped hunting them. They still kept them herded, but as protection from the other predators. They began to eat other things. It's what happens when two species discover each other."

"It didn't happen that way on Earth."

"It's because your kind only discovered more humans," Norh said. "The connotation is different. Other humans means more competition for resources. No learning is required. Had it been a different species, curiosity would have taken over."

Stephen was shaking his head, still holding the skull, staring into its empty eyes. "Sorry, but I can't see that happening. Survival always wins over curiosity."

"These people were proof of it," said Kylac. "I'm making ways to two of the settlements."

"So am I," said Deka. "Someone has to know what happened here."

"I don't think they will," said Kylac. "A third of the planet is now out of step with time. Why didn't this happen on any of the other worlds?"

"Friend is still experimenting," Norh said.

"You think this was deliberate?" Kylac said. "What was he trying to do?"

"Find new ways to destroy solar systems," Deka said with a low growl. "New ways to kill entire civilizations."

"Deka..."

"I'll feel better when we catch up to him."

"First we have to find his scent. I'll go back to the hub and try."

The fox turned and walked for the portal back to the hub. Deka went with him.

Stephen still had the skull in his hand. He noticed Norh was watching him. Stephen became self conscious and gently set the skull down. Norh's gaze remained fixed on him, unblinking and emotionless.

Stephen turned around and around, sometimes stumbling over a few bones. The skeletons seemed as if they would reach up and pull him down at any moment. The valley was silent.

"My God..."

"You keep saying that," said the Krone. "Why?"

"What do the Krone say when they see something that blows them away?"

"Why say anything?"

Stephen blinked. Nodded. Gazed down the valley silently. Norh walked toward Stephen, stepping carefully over the bones. They still shifted, clacked, and broke under his weight. He stood next to the human and breathed on him a few times. Stephen was glad for the warmth. He was starting to enjoy Norh's breath giving him heat.

2

The next region was covered in cut and smoothed stone. To the human walking next to the Krone, it resembled some kind of temple, with arches and raised platforms and pedestals everywhere.

The resemblance only applied to the mood, though, as there did not seem to be any rhyme or reason to the layout. Arches rose from the ground, reached two paces high, and then curved around and ended in the floor. Waist-high pedestals rose up from the floor in random places, scattered around like blocks in a daycare center.

Arches going every which way, pedestals popping up everywhere—Doctor Seuss in stonework, but less whimsical. Exactly the kind of place one would expect spirits to come out and play.

The travelers moved across the stone platform, pausing to scent the tops of the pedestals. A skeleton lay on top each of them.

"Guys, what is this place?" Stephen said.

"I don't know," Deka answered.

Kylac was still scenting a pile of bones. "The people of Hypsil didn't build anything. This smells as though it's been here for hundreds of years. Norh?"

"It has been fifty-two years since I came here," said Norh. He was having a difficult time navigating, as nothing here had been set up for people his size. He stepped carefully between each pedestal and arch, trying to keep up with Stephen. "I do not recognize any of this."

Kylac was scenting the skeleton on the next pedestal. "Another Hypsil. So far I smell fifteen Hrha and twelve Hypsil."

"Can you tell how they died?" Stephen said.

"No. They're too old. Just bones. But I don't see any injury, and so far they're all infants."

Kylac turned away and faced the rest of the temple area. Arches and pedestals stretched from horizon to horizon. Random staircases led to nowhere. Walls began and ended in random places. Everything was so gloomy in the dim light of the distant star.

Deka walked ahead, climbing one of the staircases. It ended at a top step to nowhere, and he surveyed the land from this perch. It was all more of the same, a chilly and scentless wind rising from it.

"Nothing is familiar," Deka said. "Kylac, Norh... Are we sure this is Hypsil?"

"It is Hypsil," said Norh, approaching the staircase. At this height, the raptor now stood at Norh's eye level. "I recognize the valleys and mountains. The skeletons are also familiar. As for the stonework, no."

"No adults," said Deka. "No bodies littering the ground."

"No blood," Kylac said. He had just caught up to Deka and was climbing the stairs. "No traces of flesh. They were placed here long after they had rotted away."

"I don't know any species that builds this way," Deka said. "Or does this to their children."

Kylac reached the top of the stairs and stood beside Deka. Stephen reached the top step moments after Kylac, all standing at eye level with the Krone.

"I have been to every planet in the contacted universe," Norh said. "Many planets outside of it. This is not familiar to me."

"Another uncontacted species?" Kylac said. "Perhaps Rive and Friend went to an uncontacted world and managed to evacuate them here?"

Silence as they pondered this. Stephen gazed out across the land.

"It could not have been longer than a month ago," Norh said. They had become used to using human measurements, as they had become accustomed to speaking English at all times for Stephen's benefit. "Not nearly enough time to fill an entire valley with bodies."

"Or build all of this," Deka said. "What are we doing here? Kylac, did Friend or Rive walk here?"

"I haven't caught their scents yet."

"Well, keep sniffing! Stephen, look around. Anything out of the ordinary, tell us."

"Right."

Stephen turned and descended the stairs. The stone was polished and smooth. Very easy on the feet. Kylac followed him down. Deka growled, leaped off the top stair, and landed on the stone ground, killing claws raised, handclaws spread. Norh turned and began searching.

There were many skeletons to sniff. Countless arches to pass through. Many unconnected walls to examine. This place never seemed to end.

3

Instead of water, the lake was full of skeletons. The four stood on the shore, looking out over the lake of the dead. The bones were white and dry, twisted together so tight there were no discernible individuals. The lake was fed by a river, but the water there had been replaced by skeletons as well.

Kylac broke out of the trance and picked up Friend's scent again. After several minutes, Deka and Norh realized Kylac wasn't here and followed him. Stephen stayed close to Norh. The cold wind blew harder here, and the dim light from the star gave no warmth, so Norh breathed on Stephen frequently.

Kylac had stopped at a particular place and was concentrating on it. Norh walked ahead of Deka and Stephen and scented it with Kylac. It was another piece of misplaced soil. They both knew the region. Kylac began contemplating the way.

Stephen turned and faced out over the lake again. Several times he opened his mouth, but he did not speak a word.

Some time later, Kylac opened the way, and they stepped through to the next region. This one overlooked another river valley, and skeletons seemed to be trying to flow in place of the water, for quadrupedal skeletons packed the channel, all facing downriver.

There was no stone here. The short grass and tall trees were still in full bloom and thriving, even though the star was nearly gone, and there didn't seem to be a trace of moisture left in the air or the soil.

Kylac dropped to all fours and followed Friend's trail. Norh, Deka, and Stephen followed the canine along the shore. Their feet made the only sounds. Stephen looked down at the river frequently as they walked by. The skeletons seemed to be arranged as if they were swimming. Stephen wanted to ask, but speaking felt deeply disrespectful now.

Finally they reached a place where the scent ended at a portal intersection, and they sat on the shore, waiting for Kylac to calculate the way. Stephen sat next to Norh. With only his boxers to wear, he relied entirely on the Krone for heat. Stephen felt guilty. He had been in the Army, so he knew how to survive in the wilderness, in the cold, in the jungle, in the desert, and yet all his training required tools of some kind, or clothing, or the familiar landscape of Earth. All those years had seemed to be God's plan to prepare him for this, and yet none of it had been useful in the rest of the universe. None at all. He looked up at Norh, wondering if the Krone resented him for being so helpless. Stephen turned back to the river of skeletons.

The way opened. Kylac rose and walked through, the others following.

Judging by the shoreline, this region could have been the ocean, but there was no water. The ocean was full to the horizon with skeletons. Unlike the ones in the rivers, these did not seem to be arranged in any particular order. Some were face up. Some face down. Some twisted together. Some independent.

Stephen couldn't hold it in anymore. He stammered for a few words, and then everything gushed out. "What am I looking at? What the fuck is happening? What in God's name *is* this place?"

It was as if Stephen had given everyone else permission to speak. Deka's words began as a low growl, and then rose to an ear-piercing screech. Kylac cried as well, making all sorts of noises which onomatopoeia could never describe.

Norh's wings were partially unfolded, and he held his head down, making low moaning sounds. Stephen held one of the dragon's legs, leaning on it much as he had seen Kylac leaning on Deka when the raptor was in grief.

Words came out of Deka's mouth. The only intelligible ones called death upon Rive and Friend and promised he would kill both of them for the sake of the dead. Their cries did not echo. They got the feeling their voices fell off the face of the planet and into a place where even the concept of sound did not exist.

4

Stephen wasn't sure if Norh was comforting him while he grieved, or if he was comforting Norh. Norh lay on the shore, facing the still ocean. Stephen sat on the sand, leaning against Norh's neck. Norh hadn't stopped making the moaning noises since they arrived, and Stephen wished they would leave, but that felt like dishonoring the dead.

Stephen had stopped crying a while ago, but he perpetually wanted to. He hadn't cried since Brenda's funeral. He wasn't the crying type, but this...

Kylac and Deka also stared out at the ocean. The fox had tried eating some of the plants, but they had been tasteless and the most innutritious things he had ever found. The Relians both remembered this world had once been full of fruit-bearing plants. It was a planet of herbivores, so there were plenty of good plants to eat, but those were all gone. The vegetation that remained was unfamiliar, but this was definitely the planet Hypsil.

What was left of it.

They had been here all day. The only motion was the wind. Sometimes it made a whistling sound as it moved over the bones. Sometimes it rustled plants. It was so quiet they thought they could hear the planet hurtling through space.

The weak daytime star set, and yet the light did not change. Deka and Kylac fell asleep. Norh and Stephen were still awake. The Krone stared out at the skeletons, still grieving for them.

Stephen stood and stepped away from Norh's neck. Norh raised his head and regarded the human. Stephen took a deep breath and walked up to the shoreline. He looked down, made sure he wasn't about to step on something sharp, and walked out onto the ocean.

He walked several dozen paces, and then he stopped, surrounded by the skeletons. He thought he'd seen visions of planets that were hellish before, but this place was as close to hell as Stephen could imagine, or perhaps what hell would leave behind if it had to move to a new location. Stephen faced the horizon, raised his hand, and shouted loud enough for everyone to hear.

"O you dry bones, hear the word of the Lord! I will cause breath to enter into you, and ye shall live, and I will lay sinews upon you, and will bring up flesh upon you, and cover you with skin, and put breath in you, and ye shall live; and ye shall know that I am the Lord! Come from the four winds, O breath, and breathe upon these slain, that they may live!"

The wind was calm and quiet. The skeletons were still. Stephen clenched his teeth and called out again.

"Is anybody out there? Anybody! Somebody make some noise! Somebody tell us what happened! Everybody can't be dead! Something must still be alive on this goddamn planet!"

Stephen's words did not echo.

Moments later, he heard crackling and shifting and breaking bones behind him. Norh stood beside him, head down to Stephen's eye level.

"What was that?" Norh asked.

"One of Brenda's favorite Bible passages. Pastor said words have power, especially the words God Himself spoke. I never really believed it. Now I know I don't."

Norh walked in front of Stephen, turned and faced him, his entire body aimed directly at the human. Whenever Norh did this, it always made Stephen feel like the center of the universe.

"Stephen, take this emotion and try to imagine yourself as the cause."

"Me?"

"Yes, you! You standing here, feeling this way, knowing you caused it. Because of you, all of this happened."

Stephen turned away and faced the skeletons around him. He did imagine it. It was easy now, somehow. He imagined himself speaking a word that made all of this happen, leaving only himself alive on this planet. He wiped a few tears away.

"Don't hold back!" Norh said. "You caused this! This is your fault! You are guilty! The blood of an entire species! Everything is gone because of you!"

Norh's words had incredible power, and for a few brief moments Stephen believed it. He actually believed it was true. Stephen cried so hard he dropped to his knees, cutting one of them on a vertebrae. Norh moved over him, touching his back with a hand. Stephen didn't know how much time had passed. He thought he blacked out for a few minutes just grieving, purging every sad event he had ever witnessed but had never expressed, all the while believing he caused it. It was all his fault. This belonged to him, and now he had to live the rest of his life with it.

Slowly he came back. He remembered he had done nothing. He could not have done anything. Norh lifted his hand and let Stephen rise to his feet. His nose took him to Stephen's bleeding knee, and he licked it until the blood stopped. He met Stephen's eyes.

The human felt so drained he could barely speak. "How did you do that?"

"Words do have power, Stephen, but they move people, not the dead. What you just felt is how the Krone felt when we realized what happened to our companion race. The Krone still live with it every day."

"My God..." Stephen shook his head, but it didn't help. "How did you do that to me?"

"Your brain was searching for an explanation. The mind takes in chaos, and it has to assign order to it. Even if the order is incorrect, it is better than chaos. Your mind wanted someone to explain it, so I provided one, and your mind latched onto it even though it knew it could not be true."

"We tell stories of chants and phrases that can raise the dead, move mountains, make it rain, make people obey... That was magic right there, what you just did."

"Another expression of your kind's lack of control over your environment. If only your words could move nature as easily as they can move people."

Stephen took a deep breath, faced Norh. "If only."

Norh lay on the bones, still with his entire body focused on the human.

"Some time after I realized the people of the volcanic planet were extinct, I found a new uncontacted world. This one was inhabited by herbivores. They were already intelligent, and not primitive. They only ate one species of plant, and it only grew in one region on the entire planet. There wasn't enough space for all of them, and the herbivores fought over territory to grow their crops.

"They invented means of killing each other. One built a weapon, another would build a stronger weapon to counter it, and so forth. So many lives lost over the right to control the food. A caste system developed. Those who owned the food, and those who grew it.

"I approached them and learned their language. Then I established myself as their common enemy. I hunted them, which encouraged them to aim their weapons at me. Later I flew into the clouds and stirred up the air, forcing the clouds to break up. It caused a drought that killed much of their food supply."

"You starved them?"

"It worked. They stopped fighting each other and began to cooperate. I hoped it meant they would realize they weren't different, that they really were living on the edge of survival, and they didn't need to compete for food. They could cooperate for it instead, and that would raise everyone higher."

"What happened?"

"They cooperated for food. I had changed that. Then they started fighting over the right to mate. The same caste system emerged, only now females were owned instead. It erupted into a war. Nothing was left."

"That's..."

"It was meaningless. Absolutely futile. I told them so. I tried to show them. All of that effort, decades I tried to guide them, and they still died. Without a companion, they could not understand."

Stephen was breathless. When Norh told a story, no matter how mundane, it was always larger than life. "You can change the weather. You can put yourself between two sides of a war and make both of them afraid. I can't imagine having that much power."

Norh raised his head above Stephen's. "I have no power."

"You sure moved me a minute ago."

Norh's eyes scanned the shore.

"The Relians are listening. They wonder when I will take you somewhere else instead of following them."

Stephen looked at them. "Good question."

"The disaster moved me to action. I wanted to help them catch up to Friend. I also wanted to help you."

"You've helped me more than I can ever repay."

"I am trying to give you a new perspective. All of this is preparation."

"It's practice?"

"You'll understand when you're ready to leave them. Try to sleep. I won't be ready to open the way until tomorrow."

Norh began walking away. Stephen kept pace beside him, hopping over the sharp bones and stepping only on the harmless ones. The wind blew across the ocean, carrying a dry scent of things that had long ago decayed.

Cham

I

The usual sight greeted them as they stepped through the portal: a dim star, air bleeding off the planet, the land cut away, columns of it scattered around, some floating in orbit or just beyond the horizon.

The hub was empty save for a short line of portals leading to different parts of the planet. Bipedal mammals were everywhere, running between the portals, shouting, screaming, searching for people they knew, asking if they had smelled this person or that person. Many had noticed the group's arrival and were running to meet them.

Norh lowered his head and whispered to Stephen. "Do not touch these people. Both the red and black species have quills on their bodies. Not venomous, but they hurt going in, and they will hurt even worse coming out."

By now the people had come close enough to study. They were vaguely humanoid in appearance, but much rounder. Their bodies were covered in fur, some red, some black, and they had long, bushy spines on their backs.

"Porcupines and hedgehogs," Stephen said.

"And these are the sentient species," Norh said. "Most of the animals have thorns, spines, or quills."

The group was upon them, scenting everyone, giving special attention to Stephen and even to Norh. Everyone spoke at once.

"A Krone? What brings you here?"

"Malim is still alive! He's evacuating us to Hezgre."

"Who is the furless one?"

"Relians! I smelled two other Relians here earlier."

"Someone tried to stop them and hold them."

Deka and Kylac kept their distance, and the natives kept theirs. It was a very orderly group, with everyone giving one another a great deal of personal space, no jostling about, no pushing or shoving.

"Where is Malim?" Deka asked in the Cham language. "Do you need help bringing survivors through?"

"Malim is on the fourth continent," one of them answered.

As one, the people pointed with their muzzles toward a particular portal just a few paces away.

"When the disaster struck," someone else said, "the portals around Cham went out. All the settlements were isolated. Malim has been bringing everyone back to the hub."

"We'll find him," Deka said. "Get off this planet as fast as possible."

"Did anyone see Rive leave?" Kylac addressed the people. "Did the exit portal cross the ground?"

"It did!" someone said. "I saw!"

"Perfect!" Kylac said. "Take me there, please!"

She moved through the group of quilled creatures, but she didn't have to nudge anyone out of the way, as everyone had their own space.

Kylac turned to Deka. "I'll go with her to find Rive's portal. You find Malim and help him evacuate everyone."

He began to walk away. Deka grabbed his tail and yanked him back.

"I'm not leaving you alone."

Kylac's tail wagged in Deka's hand. "Deka, I'll be fine."

"The stress is making you prone to reverting. Stay where you can be around my scent."

Kylac's tail stopped wagging. "I haven't reverted in months. I've been away from you before, and I was fine."

"It's not worth the risk. Especially now that—"

Deka turned and faced a familiar scent. She had made her way through, and now she stood just a pace from them. She had black fur and black spines. Everything on her was black, and in the dim light of the dying star, she lacked facial features of any kind.

"Sonjaa," said the raptor.

Her quills rose. "How do you know that name?"

Deka stepped closer. Now she picked up Deka's scent. *"I... I think I know you."*

Deka cheeped like a hatchling and moved in closer to her. He nuzzled her neck with his snout. He would have wrapped his neck around hers, but her sharp spines were in the way.

"Sonjaa... You're going to die."

Her quills flattened against her back, then rose slowly. "I don't—"

"We know each other. We had eggs together. You had a fox. Everything you remember is true. Stay with me and please if you see a crowd do not go near it."

Just to their side, some of the Cham were still around Stephen, scenting him up close, giving him no personal space. Some were speaking to him, and Norh was translating. The group had broken up, with many filing into the portal to Hezgre.

The Cham with Sonjaa's scent held Deka's snout. "This... This does feel familiar. How do you know I remember laying a Relian's eggs?"

Some of the people in the crowd were murmuring that Malim was on the the fourth continent. There were only three portals and not nearly enough scents at the hub for everyone to have been evacuated. He needed help, and time was running out.

"Deka, I'm confused right now," Sonjaa said. "I'd put these thoughts away for years."

"I know," Deka said. "You remember being on Rel. You remember being in a crowd of Relians, and you've had an affinity for them all your life."

"Yes, and the next thing I know, I'm here on Cham. I lived a life here. I studied language. Reptilian languages, mostly."

Deka backed away, held her hand and rubbed her tiny claws with his. "You are a Relian reptile, and you are about to die."

"Deka, I don't believe that. The danger is over. We're evacuating."

"Trust me, Sonjaa. You will die unless I can figure out a way to stop it."

She reached out and felt his face with her fingers. She felt him extensively. "I always wondered why I found Relian scales so beautiful. Reptile scent... I never took a partner of my own kind. Never wanted to. Now I know why."

Deka wanted to hold her hand, but the back of it was covered in short spines.

Kylac cleared his throat. Sonjaa and Deka turned to him. "We need to find Rive's portal site, as well as Malim."

"Yes," Deka said. "Sonjaa, you saw them leave?"

"I'll take you there."

She dropped her hand and began walking out of the trees and into the wind. Deka walked beside her, as close as he could be without hurting himself.

2

The wind was calm here, and it was dark. Some of the moons reflected a little light off the dayside of the planet, but they did not help much. Deka, Sonjaa, and Kylac needed night vision to see. They were the only three people on the continent.

"I remember you told me about the first time you came here," Sonjaa said. "Kylac wanted to see people covered in quills, so you followed the fox all around Rel, looking for the portal to Cham.

"You became fascinated by the locals. You had just started hunting for your fox, and you wanted to bring back challenging prey for a change. So you took off after one of the predators. The next thing you knew you had spines in your muzzle, and you were on your back, screaming. Kylac tried pulling them out, but they were barbed on the end."

Deka was rubbing his claws while they walked. "I had no idea animals here could throw their quills. Kylac tried to be gentle, but there was no way to work the spines out without causing more pain. Thankfully for us, a few Cham saw what happened, and they knew how to work them free. They had to dig their claws into my muzzle and part the scales, and the quills slid out."

"Just another young predator," Kylac said, "taking on prey he was never adapted to catch."

"During the surgery," said Sonjaa over her shoulder, "you managed to have sex with one male and two females. Even the Cham were distracted by your scent."

Kylac's tail waved. "Is it doing anything for you?"

Her quills rose and fell. "None. I've been told the canines of Rel have that effect on everyone, but when I visited, I felt nothing. Now I understand why."

Deka took the risk of walking up against her. Her quills were just one wrong twitch of the hips away from spearing him.

"I had to try," Kylac said. "So where's the portal?"

"Rive and Friend were here with Malim. I saw them talking together over this den."

They came to a large hole in the ground. Nobody was here now, but the scents of hundreds of Cham led away from it to the portal which led to the hub. Kylac scented it as they walked by. Definitely evacuated.

"Did you hear what they were talking about?" Deka asked.

"I didn't hear most of it, and the rest I didn't understand. But... Friend grabbed Malim and yelled at him to control himself. He said everything he had done since the disaster was to fix what he destroyed."

Deka huffed. "There's only one way to fix that."

"Not too long after that, something happened. Our star was cut in half, and so was most of the planet. Friend was shaking Malim, still yelling at him."

Deka and Kylac met each other's eyes. They kept walking for some time, and then Kylac smelled something that did not belong. He ran ahead of them, dropping to all fours and scenting the ground.

Finally he raised his head and faced Deka. "You're not going to believe this."

"What?"

"They went to Vico."

"Are you sure?"

"Yes."

Deka broke away from Sonjaa, caught up to Kylac, scented the discolored patch of ground himself.

"Vico..." Sonjaa echoed. "Isn't that one of Rel's moons?"

"It was. I'll make the way," Kylac said.

"Why did he go to Vico?" said Sonjaa. "Rel is gone. The moons have probably scattered."

"They have by now," Kylac said. "It will take some time to calculate. That's a lot of variables to work out. This won't be easy."

Sonjaa's quills flattened against her back. "That's why I stuck to language. Thinking about physics all day eats up your whole life, but language at least lets you interact with people."

"Let's go to the fourth continent," Deka said. "We'll find Malim and ask him what happened."

Kylac stood up, and they walked back to the portal together. Then Kylac started running, pulling away from Deka and Sonjaa.

"Kylac?" Deka said.

"I think Malim is in trouble!"

Deka took off after Kylac. Sonjaa ran as well, and though she could never match a raptor's pace, every muscle in her body tried.

3

Stephen stood off to the side, taking in the planet of Hezgre, a strange world full of vines and shrubs but nothing that resembled a tree. The Hez were serpents about as large as the anacondas Stephen had only heard about, though their heads resembled Norh's more than a snake's. Their companion species was a race of equines, the Gre. The entire hub was crawling with snakes and horses.

Quilled creatures poured out of the portal and began making themselves at home. Now that Stephen saw them in better light, he realized they resembled hedgehogs and porcupines only in the length and shape of their quills. The people of Hezgre met with the quilled refugees and guided them to the lake. Some had begun digging burrows.

Norh walked up beside Stephen and nudged him with his head. Stephen stumbled forward and turned to the Krone. Norh's wings were partially unfolded. He was in a good mood.

"Who's telling these people what to do?" Stephen asked. "How do they coordinate all of this?"

"What do you mean, coordinate?"

"Well, look at them. Who's organizing where they should dig? Who's assigning people to help the refugees? Who's arranging where they drink, where they eat? How does everyone know what to do? I've noticed it with every planet I visit. Where's the government? Who's in charge?"

"Why should someone be in charge?"

"Someone has to coordinate, or nothing will get done. Everyone will bring potato salad to the potluck."

"Why would anyone not know what to do?"

Stephen sighed and watched the refugees coming in. They were definitely people who respected each other's personal space, which seemed to be in contrast to the Hez, who crawled all over everyone and everything. It was disconcerting to see so many anacondas in one place.

"Don't you teach your people how to live?" Norh said.

"Of course we do."

"Then why do you need people to control other people?"

Stephen held up a hand. "Forget I asked."

"I think this is a good question. Humans need some sort of authority over them for order to be maintained. Why can they not control themselves?"

"Just forget it."

"It maintains order, but why do you need someone to maintain order? It's because you are still locked in social systems that favor survival of the fittest. Government is the only thing keeping you from slaughtering one another in a literal sense. It is another symptom of a species that lacks perspective."

"You know, this is getting insulting."

"To you, every other human is still a threat to your survival. You know this on a subconscious level, and it affects how you treat one another. You cannot accomplish anything as a species until you learn your prosperity does not require someone else's demise. That will not happen until —"

"I know, I know! It wouldn't happen if we had a companion species. The root of all our problems. I know."

Norh lay on his stomach, keeping his head at Stephen's eye level. They watched the Cham and the Hezgre for a few minutes.

"What's the point of seeing all of this?" Stephen said at last. "I'm meeting aliens, but I don't understand half of what I'm seeing."

"And yet you want to meet more."

"I wouldn't have passed up the chance for a billion dollars. I don't care if I look for a hundred years and still don't understand. I'd still spend my whole life looking at them. Maybe if I did it enough, something would finally click."

Norh was silent. Stephen was sure he would have something to say to that. He wanted the Krone to tell him there is hope, or there is a way, or this is only the beginning. Instead the Krone said nothing.

"Maybe I can emigrate," said the human. "Learn a language, settle somewhere else. I sure as hell don't wanna go back to Earth, not after seeing all this. The universe is too big to waste my life in a factory. For the first time, I have actually lived, and I never want to do anything else."

"If you could, with whom would you live?"

He sighed. "I'm not sure. I've seen so many people, and it's been shoved in my face a million times that I can't survive anywhere. I don't know where I can go. I wish I could go to Earth and make them understand. The world is crap. Everyone knows it, but nobody knows what to do about it. I wish I could show them what to do."

"There may be hope for you, at least."

Stephen turned to Norh and met his eyes. He smiled.

Norh's wings were unfolded slightly. He was smiling back.

An anaconda-thing with a dragon head crawled up Stephen's leg and began wrapping itself around his waist. Stephen did not feel any reflex, and Norh didn't need to tell him this individual was just saying hello.

4

There were far too many people in the settlement the Relians crossed into. They were on the fourth continent, far away from the disaster, but everyone had felt the planet tear apart.

Some saw the portal open and rushed to meet them.

"They're here!" someone called and ran down the closest burrow. Moments later, hundreds of quilled people poured out and ran to meet Deka and Kylac. They formed a line and began crossing through the portal to the hub, re-uniting with people from the other continents, and then moving to the safety of Hezgre.

"Malim isn't here either," said Sonjaa. "This was the last place on the fourth continent. Where is he?"

"Kylac, I hope you're wrong," Deka said.

"So do I, but we've been everywhere on the fourth continent, and Malim still hasn't evacuated it."

There were no wounded here, so the Cham passed through the portal completely on their own. The line was panicked but still orderly, with everyone moving at a brisk but measured pace.

Deka waited for a break in the line and then ran through the portal himself, Sonjaa and Kylac following. Back at the hub, Deka walked away from the people and prepared the way to the next continent.

Sonjaa sat next to him, leaned against him. Her mammalian body was warm, even in the cold wind rushing off the face of the planet. He leaned into her, wishing he could embrace her as a Relian.

"I wish I'd kept trimming my quills," she said.

"You kept them cut? Why?"

"To help me fit in with the reptiles. But I haven't had them cut in years—I was tired of keeping up with it. I've been told I acted like a Relian in how I expressed affection."

Deka nipped her lightly on the nose. "So why didn't you live there?"

"I did for a few years, before the disaster."

"When was that?"

"I was on Rel six years ago. Lived there for five years. I came back because I didn't find what I wanted."

"You were searching for me?"

Her quills raised and lowered.

"And in all that time, you never caught my scent?"

"Not that I remember."

"There were only four Archeons on Rel. I was everywhere. You must have met me."

"No, I never did."

"And a Cham without her quills, living on Rel as a raptor. That would have stood out. Someone would have mentioned you."

"People did talk about me. I was an oddity there. People found me unusually eager to be close to others, and since I wasn't affected by fox scent, people joked I was a raptor."

Deka stared at her. "This doesn't make sense. You were on Rel for that long, and we never met? What did you do all those years?"

"I learned the languages of many reptilian species who came to Rel. I visited other worlds with reptiles. I lived with them, too. Even hunted with them." Her quills rose and fell a few times, chuckling. "The other raptors on Rel enjoyed seeing me hunt."

Deka tilted his head. A few breaths later, a way opened to a settlement on the fifth continent. Deka stepped through and Sonjaa followed close behind. Kylac stood still for a breath, tail flicking from side to side, then he walked through.

A sea of people waited here, black quills and red quills. They surged toward the Archeons. Sonjaa explained where to find the portal to Hezgre. Word traveled fast, and they began filing through the portal, frantic, but still evenly spaced.

Kylac waited until everyone was out, then he crawled down the hole. Deka sat with Sonjaa. Surrounding them were trees and short grass, and various animals all huddled together in a trance, all of them covered in quills of various lengths and shapes and positions.

"I have needed you all my life, Deka," Sonjaa said. "That's how I felt since the day I hatched. Like I lost someone in my family and I've been grieving for him for years. I can't explain any of it, but I've spent my whole life trying to find you again."

"Your whole life? Sonjaa, the disaster that destroyed Rel was only a year and a half ago."

"I know. I hatched thirty-four years ago here. That's... twenty-three Relian years, isn't it?"

"Twenty-seven."

"Things became more real after the portals went out. The memories became stronger, and when I couldn't leave Cham, all I could think about was trying even harder to find you. It's been very frustrating."

"For me, too. You're the fifteenth incarnation of Sonjaa I've met since the first disaster. I have watched Sonjaa die fourteen times right in front of me. I will do everything I can think of to keep you alive, but..."

Deka whimpered. He reached over and nipped her nose. She nipped his.

Kylac came running up to them, then slid to a stop and rose to his hind legs. "I searched the burrows. Malim was never here. I think I know where he might be."

Deka stood, and so did Sonjaa, still as close to him as possible without hurting him.

"Where?" Sonjaa asked.

"We should go back to the first settlement he evacuated. I think he'll be there."

"What makes you say that?" Deka asked.

"He will want to be away from everyone."

"We can't go there now," Deka said. "We still have eight more cities to evacuate."

"You evacuate them. I'll go—"

"No, I'm not leaving you alone, Kylac!"

The fur on the fox's back rose. "Deka, finding Malim is our priority! These people have another day or two before the planet is uninhabitable. If we don't find Malim, there will be another disaster!"

5

The way opened, and two Relians and one black-Cham hopped out. The wind was strong, and the animals were still huddled into themselves. Predator lay with prey, comforting each other, as if realizing nothing mattered anymore.

Kylac ran to the hole in the ground and scented around it. Deka bent down and scented the dirt as he walked in Kylac's footsteps.

"I hope Kylac is wrong," Sonjaa said.

"It all adds up. Kylac can smell Malim's mental state in his scent better than I can, and I agree. Everything we've found so far points to it. I trust him."

"You always did," she said. "Even when you were young. Relians are so strange, how the raptors and foxes have grown to the point where they need each other. Raptors are family-oriented. Foxes are always pushing raptors to leave the family and do other things. But the foxes are savages without their raptors, so the raptors are always pushing the foxes to use their higher minds. It's poetic. It really is." She lowered her quills and her voice. "I sought a fox once."

"You did?"

"A year before I left Rel, I met a fox named Behee. She had only been born a year before, had no raptor, and we shared a connection. We were together for a while. And I thought... I thought I finally knew how the raptors felt."

Deka stared at the face he couldn't see in the dim light. "Tell me."

"I don't know if I can. When I was with her, I never wanted to be away from her. I wanted to be with her for the rest of my life. I wanted to help her, and she wanted the same thing. Then one day she met a raptor, and now Behee was with her all the time. I still saw her sometimes, but the bond was gone. I missed it."

"I'm sorry. It usually doesn't happen that way. When a raptor finds a fox, there's usually no changing it, but since you're not a Relian..."

"It was a reminder that I'm a Cham. That's why I came back. I knew I wouldn't find what I wanted."

Deka shook his head. "Something like that... It would have been news all over Rel. I would have heard about it. I should have met you—Why didn't I?" He paused, turned to her again. "Sonjaa... How did you die? The first time, I mean? On Rel, how did you die?"

"I've been asking myself what happened since I hatched. I don't know what happened."

"Deka! Sonjaa!" Kylac was shouted at them from far ahead. "He's here! The scent is fresh!"

Deka took off running after the fox. Sonjaa ran as fast as she could, trying to keep up. Deka quickly caught up to Kylac and kept pace with him as they ran straight for the jagged horizon, where the planet had been cut off.

A dark figure stood barely visible against the dim sky. His quills lay flat against his back. As they approached, they saw he was perched on the edge of the planet, looking down.

Kylac and Deka stopped a few paces from the edge. The wind was too strong to feel comfortable venturing closer. Sonjaa caught up a few moments later.

Kylac spoke first. "Malim?"

The red-Cham turned and faced Kylac. "You found me."

"This is where I would come if I didn't want to hurt anyone with a thought I couldn't get out of my head."

His quills remained firmly pressed against his back. "You know what happened?"

"Sonjaa told us," said the fox. "What did Friend tell you?"

Malim looked out over the edge of the planet, perched precariously on it, as if he did not care if he fell or not. "He told me everything. He gave me coordinates, variables, and a new way of perceiving the universe. Then something horrible happened. I opened a portal I couldn't stop."

Deka swallowed.

Kylac took a step forward. "Malim, please. Tell us what's happening. What is Friend doing?"

Malim's quills rose, fanning all the way up. "You don't want to know! If you do, you'll be like me! You can't stop thinking about it! This is a virus infecting thought itself, and now... I caused the disaster on this planet! It wasn't Friend or Rive."

"We know," Kylac said. "What is Friend trying to do? Why can't he control it?"

"Friend wouldn't say, but he thought I could handle it. He was wrong. As soon as the antisphere appeared, he and Rive left. I managed to shut it down before it swallowed the planet by going to different parts of the planet, but I'm having a hard time focusing on the portals. I only got to the fourth continent before I lost them completely. The new equations pushed the portals out of my head. Then another antisphere opened and it cut off even more of the planet. That's when I came here."

Kylac took another step toward him. "Come back with us, Malim. Let the Selts examine you. Maybe they can help you figure out what's wrong."

"I'll destroy Selta, too! I'm just like Friend now. Everywhere I go, I'll destroy everything. I grieve for Rel even more than I did when I heard the news, and now I grieve for Friend. Friend is right! He has pulled away another layer of the universe! When a conscious mind tries to perceive it, it changes the universe itself. That's what the antispheres are. That's what's happening to all the planets he's destroyed. I can't live like this. I've killed so many people already."

"Malim, don't talk that way."

"Deka," said the red-quilled Cham. "Kill me."

Deka raised his neck.

"I can't do it myself, but trust me," the Cham continued. "If you kill me, you will prevent the destruction of every planet I visit."

"You're sure you want to do this?" said the raptor.

"Deka!" Kylac glared at him.

"Kylac, Friend destroys half the worlds he visits! If Malim can't—"

As they spoke, something appeared within the space already cut away by the disaster. A sphere that led to nowhere. The antisphere grew slowly but ominously.

Sonjaa broke away from Deka and approached the ledge, her quills blowing toward the cliff. The two Relians backed away.

"This has been happening since the disaster!" Malim screamed. "I can't stop them from forming! This is the safest place. I can control where they open, but only so far. I can control the speed they open, but not the size. This part of the planet is already destroyed. I won't hurt anyone here."

"What do you sense?" Sonjaa shouted.

"That is time itself," said Malim. "I understand it now. Separating time from space. He's right. Friend is absolutely right. This is where the universe is expanding into—where it's moving. Something exists outside our universe! This is the proof!"

The antisphere grew. The ground began to shake. Deka and Kylac were backing away, the raptor looking at the fox, silently telling his fox that Malim needed to die, the fox silently telling him they needed more information.

Sonjaa braced herself on the ledge with Malim. "Time... I was on Rel. With Friend. I was trying to help him! Deka, I remember! It's happening again!" She turned and faced Deka as the antisphere grew closer and closer to the ledge. "That's what I was doing! Deka, I was with Friend! I was trying to help him stop the portal! It was out of control, and I knew it wasn't a normal way! Malim!"

She jumped the three paces over to him, grabbed his waist, and turned him to face her. "Look at me!"

"Kylac," Deka said, "get ready to grab me in case I fall."

"Deka, no! If we kill him, we'll never understand what's happening."

"Maybe that's a good thing." Deka crouched, spread his hands, raised his toe claws. "Sonjaa, get away from him."

"I can help him! Malim, remember when I—" The wind swept her next words off the ledge.

The antisphere loomed large just beyond the horizon. The planet had already been ripped up, so there was no tearing or destruction. The antisphere grew larger and closer. Sonjaa reached up and grabbed Malim's face, forcing him to look at her. Deka and Kylac couldn't hear.

Deka kicked away, claws raised and ready. As soon as he took the first step, he slammed into another raptor. Deka halted, stunned. Suddenly the land was full of Relians with some other species mixed in. Their scents hit Deka—the same scents he had smelled on Faii.

The crowd around him was thick, with people gawking at the antisphere in the sky slowly spreading toward them. Unlike a normal sphere, its destination was not projected around its surface, and it seemed to project nothing onto it. Perceiving nothing was the most hypnotizing feeling Deka and Kylac had ever felt.

Deka looked at Kylac. He, too, found himself in the midst of hundreds of Relians that had suddenly appeared. He was just as confused as Deka.

Sonjaa had taken Malim down to his knees, still holding his face in her hands, shouting to him.

"Sonjaa!" Deka tried to push through, but nobody would move. He tried to jump but had no room to bend his legs. The crowd pressed in on him tighter and tighter.

Deka turned his head and looked over his shoulder. The landscape was full of raptors and foxes, all standing like stone and staring at the antisphere. Deka tried to back away, but the fox behind him stood in a trance. Deka snarled and clawed the fox down the chest. The fox did not react. Deka clawed him again, and still the fox did not move. Then Deka realized he had drawn no blood. His claws had not even disturbed the fur.

Kylac was trying to move as well, and he had managed to raise a hand and was clawing everyone he could reach, but his claws did nothing.

"Sonjaa!" Deka screamed. "Get out of there! Get out!"

Sonjaa had forced Malim to the ground. The wind stole her words. The antisphere had grown up to the ledge, filling most of the sky. Deka pushed and shoved, but these people were solid as planets, unblinking as they stared into the antisphere.

Malim snapped his head away from Sonjaa and ran, leaving Sonjaa perched on the ledge. The antisphere hit the cliff and ate it up. Sonjaa rose into the sphere, floated, turned around six times.

"Deka, help me!" she screamed, her voice coming from everywhere.

Deka growled and snarled and clawed and slashed at the immobile people around him. Kylac bit people's necks, but his teeth didn't penetrate any flesh or even dent the scales or fur.

"Deka, what's happening?!"

Sonjaa was flying toward the center of the sphere. Behind her, the antisphere ate up the ledge in large columns.

"Friend! Rive! Help me, please! Someone! Help!"

Before she reached the center, she divided into columns and dissipated into the emptiness. More of the land ripped up in columns, disappeared into the sphere, and vanished somewhere inside. The statue people around Deka and Kylac ripped apart in columns as well, and the Archeons wiggled and thrashed but could not free themselves.

The antisphere evaporated. The remaining statues around Deka and Kylac flickered away. The two Relians stood on the ledge, now a dozen paces closer to them than it had been. On the edge of the planet lay a bloody, disfigured Cham. Deka and Kylac ran up to it. Malim had been torn apart. Most of his head was gone.

Deka snarled, dropped to his side, and mentally crawled back into the egg, crying. Kylac lay on top of him, whimpering.

6

The portal to Hezgre had collapsed when Malim died. Deka and Kylac nearly panicked when they realized Norh and Stephen were nowhere to be found, but they hoped they would realize what had happened, and Norh would make another way, as Deka and Kylac did not think either of them could hold an offworld sphere open long enough to evacuate anyone. In the meantime, the Relians gathered the last of the Cham at the hub to wait.

As expected, a little more than a day later, a new portal from Hezgre opened right where the old one had been. The people waited for the Archeon to step through first, and Norh did, followed by Stephen, who had a large serpent wrapped around his upper torso. The Krone stood before everyone, folded and unfolded his wings, then stepped aside, and the last of the Cham began leaving their homeworld.

Stephen walked up to them. His boxers were even more tattered than before, but he still wore them.

"What happened, guys?"

Deka crouched low to the ground, eyes locked in a hunting stare, but without prey in sight. Kylac stood beside him, one hand on the back of his neck.

"I'm going to kill him," Deka said, in English. "No negotiation. No waiting. I will kill him."

Stephen backed away, looking at Kylac. The Hez coiled around Stephen also reacted with fear.

"Norh," Kylac said, "I'm about to make a way to Vico. I want you to come, too. I have a feeling this is it."

"Friend is close?"

Kylac's posture fell. "Very."

"I will make a way."

"Take Stephen somewhere else. It's not safe for anyone."

Deka growled. Kylac rubbed his neck harder. Deka growled louder. Stephen backed away, stood beside Norh.

Vico

Deka and Kylac cautiously stepped through the way onto a tiny moon with a thin atmosphere. Two more portals were open about thirty paces away from them, each hovering five paces above their heads, the lower atmosphere of a planet projected around them. Both spheres were spinning, pouring air in all directions.

The Relians had done something like this as part of their Archeon training. Gravity was light, rotational speed was quick, there was no atmosphere. This place was a challenge, and it was a good world to practice spinning portals in atmospheres. Kylac had figured Rive would be at the same place all of them had practiced as apprentices, so he had set them down close to it.

Friend knelt thirty paces away, directly underneath the two portals that supplied air. Rive stood at a distance so as not to break his fox's concentration.

Deka snarled.

"Deka, wait!" Kylac whispered as loud as he could. "We have to figure this out!"

Deka raised his claws and took off. His charge was awkward in such weak gravity, but Deka adjusted his stride and began moving as though he'd been hatched and raised on this world.

A raptor with light brown scales heard him and turned. The grey metal coiling his body like an affectionate Hez replaced the skin and limbs he had lost when Friend's first antisphere ripped him apart. The flexible metal reflected a distorted view of the moonscape and Deka's snarling muzzle. Rive jumped in front of Deka and crouched, claws ready—one set a single piece of metal formed in the shape of an arm and fingers, the other real. Deka leaped, trying to jump over him and reach the fox. Rive reached up with one arm, grabbed Deka by the ankle and threw him to the ground, holding him in place.

"Deka, stop!" Rive shouted, his voice metallic. "Friend is about to—"

Deka slashed Rive, but his claws only slid off Rive's metallic underside. Deka snarled as Rive held him down.

Meanwhile, Kylac had circled around and now stood in front of Friend. The older fox's fur had grown back for the most part, but his tail was still missing.

Friend opened one eye. "Hello again, you two."

"Kill him, Kylac!" Deka screamed as he tried to find something on Rive he could hurt. The metallic raptor held him down with seemingly no effort.

"What are these broken spheres we keep finding?" Kylac said.

"I don't know," Friend replied.

"What happened to Sonjaa?"

"I don't know."

"What happened to Hypsil?"

"I wish I knew."

Kylac clenched his teeth. "*Why* don't you know?"

"If I had answers, I wouldn't be struggling right now. Everything I've done has been leading up to this. You arrived just in time. I know Rive has been leaving a trail."

Rive turned and glared at his fox. "Friend..."

"You have every reason to be afraid of me. I didn't think anyone but these two would ever follow us."

"That's bullshit!" Deka said, spitting the English word. "You kept Rive around because you needed him to make your escapes!"

"That's true," Friend said. The calm in his voice contrasted with the turmoil in his scent. "I can't make ways normally anymore. This is all I can think about. I've been trying to find a way to fix the damage I caused."

Kylac stood still, staring Friend down. Deka watched him from underneath Rive, growling, trying to urge his fox to attack. Kylac stood perfectly still just ten paces from Friend, turning his gaze to Deka, then to Rive, trying to tell Deka an attack would be unwise.

Friend continued. "I've been trying to perceive the motion of the universe through time and make a way through that motion. I'm testing a new theory, that a conscious mind can't pass through time. It's the opposite of traditional portals. So I've been practicing the opposite. Trying to pull other things through."

Deka snarled and struggled again and called to Kylac to attack. Rive held Deka down, watching Kylac, silently hunting him. Kylac paced along an arc in front of Friend, keeping his distance.

"Is that what you did to Hypsil?" Kylac said. "Did you try to restore it?"

"I tried, but I couldn't find Hypsil in the past. It wasn't where it should have been. All that came through was what you saw."

"And what did you tell Malim?"

"I thought he could handle it. He's always had a solid mind, and the Cham are some of the most stable-minded people I know. Is he all right?"

"He's dead. Killed by his own antisphere."

Deka bucked and squirmed and snarled. Rive remained an iron sculpture holding him down.

"I know how he felt," said Friend. "I don't sleep. I barely eat. This thing in my head won't be silent. I have to think about it. I have to work it out. It must be solved, and now I'm ready to solve it."

"Kylac, kill him! Kill him!" Deka's words degenerated into mindless snarls and screeches.

Kylac remained still, staring down at the kneeling fox under the spinning portals. "What are you doing now?"

"You'll find out. I'm almost ready."

"You know I can't let you do that," Kylac said. "How many planets have you destroyed?"

"Fifty-six. I remember every one of them."

"Someone has to stop you, Friend."

"Many tried, but when someone breaks my concentration, I lose control, and an antisphere destroys their world."

"No loss here," said Kylac.

"Wait a moment," Friend replied. "When this works, I can go back to every one of those fifty-six worlds and fix what I did. I can't bring back the people, but I can bring back their planets."

"Their—?" Kylac said. "Vico... You're going to bring back Rel?"

He had no tail, so Friend rubbed his fingers together.

At that moment, a portal opened up just a few dozen paces behind Kylac. The portal expanded to Krone-size, and a moment later Norh stepped through, eyes fixed on Friend.

Deka snarled from underneath the metal and flesh raptor. "Now, Norh! Kill him! Kill him now!"

Kylac turned to the Krone. "Norh, wait!"

The Krone looked down at him.

"He'll succeed!"

"Kylac!" Deka screamed.

"Think about it, Deka! Every incarnation of Sonjaa said she saw you when she died! She saw me, too! This is going to work! Somehow, this will work!"

Deka's words disappeared into snarls and growls. Rive raised his metal arm and clamped Deka's muzzle shut.

"Thank you, Rive," Friend said. He closed his eyes again.

Norh's wings partially unfolded as he looked from Kylac to Friend. Through Norh's portal, Kylac could see Stephen peering in, wondering what to do.

Many breaths passed. Kylac turned around, faced skyward. He couldn't hold the offworld portal in his mind anymore, so he let it close, but before he lost the way, he remembered Rel would be right here, exactly where he was facing.

An antisphere opened far in the distance. The only way they could see it was by the stars it blocked. This time it wasn't a portal into nothing. Something was inside it.

Kylac gestured to Stephen to come through. The human stepped out, gasped in the thin air, struggled in the weak gravity, and shivered in the cold. He walked up beside Norh, and the Krone breathed on him a few times.

Deka continued struggling under Rive.

Rive was looking over his shoulder at the sky.

Norh faced where Kylac stared.

Stephen turned to where Norh's eyes pointed.

The sphere grew. It began to glow. Something molten was inside it. Everyone recognized it as the core of a planet. The layers built up, covering the core with a cooler mantle as the antisphere expanded.

Deka stopped struggling and watched.

Friend screamed and howled louder and louder as the planet grew from the inside out. The surface began to take shape. The continents formed. The mountains wiped into existence from the ground up. Then water. Then the upper atmosphere.

The sphere dissipated, leaving a complete planet in its place with continents and oceans and even a glint of starlight over its red and blue surface. Kylac reached up, as if to touch it. It had been so long since he had seen Rel from this perspective.

Friend's screaming cut off.

Everyone faced him, and the fox collapsed on his side. Rive jumped away from Deka and ran to his fox, scenting him. Deka rolled to his feet, gasping and choking as he stumbled to the fallen fox. Norh, Stephen, and Kylac approached as well. Friend was still breathing.

Rive turned to everyone, relaxing. "This is the first time he's slept since we lost Rel."

Norh swung his neck around and took in the sight of the new planet. "I'll make the way."

Rive turned his muzzle upwards. "So will I."

Stephen was too breathless staring at Rel to speak. Kylac and Deka stood still with their mouths open.

Rel

Two portals opened, one Krone-sized, the other Relian-sized. Norh and Stephen stepped through the larger portal. Through the smaller sphere walked Rive and Friend, then Deka and Kylac. Friend was barely awake, leaning on Rive as they came through.

The hub on Rel ringed the entire central continent, with the portals situated along either side of the pathway. People were everywhere—Raptors, foxes, and every other species in the contacted universe, all walking between the spheres.

But there were no portals on the hub. The people walked into the spaces where the spheres were supposed to be and disappeared into the air. Others emerged from the air and appeared in front of a place where the portal would have been. Deka and Kylac knew exactly where they were, and where each portal space should have been. They were coming from millions of light years away, and yet nothing was there.

Friend hyperventilated. He let go of the metal raptor and dropped to his knees, scented the ground. Now his body matched the panicked state of his scent. He had only been asleep long enough for Rive and Norh to calculate the portals, which had been about a quarter of a day on Rel.

They had woken him because Rive knew he would want to be awake when they first set foot on a planet he restored.

The noises Friend made as he scented around were everything but calm and assured.

"Friend?" Rive said.

"I don't know! I don't know! This wasn't supposed to happen! My calculations were perfect! Living minds are not supposed to be able to pass through! They shouldn't be here!"

"Where are the portals?" Kylac said. "Where are these people going?"

As they spoke, dozens of people walked into the empty spaces and disappeared. Others emerged from nothing and walked about. Friend looked at them. He screamed and ran on all fours toward the hub. The people on the hub did not notice him. Nobody turned their heads or slowed their pace as the canine ran up to them.

Friend leaped onto the hub path and stood on his hind legs. Relians and various other species walked by, some on four legs, some on two, a few on six. The rest of the group slowly approached the hub.

"Excuse me!" Friend said to a passerby. She ignored him and continued on her way. "Excuse me," Friend said again, this time to a male insect. It ignored him and kept on walking.

Friend stood in front of a raptor as she walked. "What are—" The raptor walked right into Friend, knocking him over. Friend became entangled in her feet for a moment as she stepped over him, never pausing or stopping. She passed over him and continued on.

By now the others had caught up to Friend and stood by the hub. Nobody noticed the Krone, or the human. Nobody even glanced their way.

Deka stepped into the path, holding his claws out. Two people walked by, and Deka's claws raked their faces.

His claws seemed to glide over air instead of flesh, and the two offworlders never flinched.

Friend had rolled to his feet and was dodging more people as they walked by him, threatening to bowl him over again. He weaved his way through, and then stood with the others just off the pathway.

"I heard somebody talking about solar flares on Faii," Deka said. "I remember that, too. That was just before the disaster."

"I don't understand," Friend said. "There shouldn't be any people here. What is this? What's happening?"

"Why don't you know?" said Deka, approaching the fox. Rive moved between them. Deka snarled. "You're the one playing with this stuff! Why don't you know what's happening? You've destroyed enough planets! You should fucking *know!*"

The human curse word blended in perfectly with the Relian language.

"It's not supposed to be this way!" Friend shouted back. "I pulled Rel through just before the disaster, but there shouldn't be any people!"

"What did you expect to happen?" Kylac asked. "You pull an entire planet from the past, what happens to the people?"

"It should not have interfered with what already happened. I don't know what's..."

"We've seen this before," said Kylac. "On Cham. These people are from the past. A past we're not part of. We can't influence them or change anything."

"Yes, yes," Friend said. He was hunched over, exhausted, his mind still racing beams of light. "That's exactly right. This is Rel, mere breaths before the disaster. I did pull it from the past, but everything is happening just as it did before, but it's not supposed to. I only wanted the planet."

Deka snarled. Rive reached out with one hand and shoved Deka backwards a few steps. Deka stood firm, crouched and looked at Friend from the waist. "If you didn't have Rive, you would be dead by now, and none of this would happen!"

"I don't think that's possible," said Friend. "Everything already happened. I only pulled the past here. The portals... They did exist in the past, and everything is happening as though they were there. Nobody will know we're here. And this planet is about to—"

Deka snarled at him, took a few steps in his direction. Rive held out a hand again. Deka stopped before he ran into it.

"How long do we have?" Kylac asked.

Norh's voice came from high above. "It is a quarter through daylight."

"The disaster happened just past midday," Deka said. "Plenty of time. Here's what we're going to do, Friend. I will find Sonjaa, and I will follow her because I want to know exactly how she died! You and your raptor are coming with us."

"I was going to suggest that," Friend said. "Sonjaa was there—"

"I know!" Deka growled. Friend winced.

Norh unfolded his wings, reached out, and grabbed Stephen. "I'm taking Stephen to see the planet before the end. I will leave this portal open. Don't stray too far from it."

Human in hand, Norh took flight. He reached the horizon in mere breaths.

"Who is that?" Rive said. "The biped, I mean."

"Uncontacted species," said Deka.

Rive held his claws together.

"I'm making a way to where Sonjaa was when I left her," Deka said. "It won't take long."

"I'm keeping a way offworld in mind," Rive said.

"So am I," said Kylac.

Friend did not speak. He was staring off into space, his mouth and tongue moving, talking to himself.

The people on the hub moved by, walking into and emerging from imaginary portals. They talked, they exchanged greetings specific to each species, exactly how the contacted universe used to be.

2

This region of the planet was hot and dry. Raptors and foxes lived here, but it was still early, so most were asleep. Deka knew exactly where he and Sonjaa had been just before he left, and he ran across the sandy soil until he saw her. She was still asleep, Rupi just a pace away from her, also asleep.

Deka had seen her in so many different bodies since the disaster he had forgotten she was a theropod. Her scales were green and brown, speckled with white. He barely noticed her scales when she was alive, but now they were all he could look at.

Their three hatchlings slept beside her, one curled up to the canine. Deka scented each of them, whimpering. He stood over Sonjaa, knelt beside her, lay a hand on her neck. He touched air. Deka pressed harder, but his hand wouldn't penetrate. Kylac stood on the other side, watching.

"Your hand never actually touches her," Kylac said. "There's a gap. And have you noticed the tiny gap between our feet and the ground? We can't interact with this world. We're not supposed to be here."

"So time travel is possible," Deka said. "Fifty-six planets destroyed just to discover this. How many lives are you responsible for, Friend—do you remember that number?"

"Yes," said the tailless fox. He stood as if he were about to collapse, asleep, at any moment, and his scent was agitated and terrified. "Two million six hundred and forty-one thousand nine hundred and three people."

"That includes all the civilizations you ended?"

"Deka," Rive said. "Stop."

"Stop. Right. Stop! It's just that easy!" Deka rose, and so did his toe claws as he turned and faced Friend. "All you have to do is stop! Stop thinking about it! Stop the portal! Stop the antisphere! Just stop—that's all you have to do! Why is that so hard?" Deka growled, still glaring at Rive. "I will stop as soon as he does!"

"Deka, this is paternal instinct talking," Friend said. "You're an Archeon. You're above this."

Deka locked eyes with Friend. He snarled, and his voice rose to an ear-piercing screech. He charged. Rive stepped between them, bracing himself. Deka slammed into Rive, and Rive shoved him backwards with his metal arm. Deka twirled and stumbled but remained on his feet, panting.

"I can't undo what I already did," Friend said, looking at the ground. "I'm sorry. I can't change the past."

Sonjaa stirred, lifted her neck and yawned. Deka saw it out the corner of one eye, and he turned to her. He never realized how much he would miss seeing her yawn. He knelt and reached out to embrace her, but his arms never touched her.

Sonjaa stood, effortlessly taking Deka up with her. He let go, and he watched as she nudged each hatchling awake. As they yawned and cheeped, Sonjaa poked her fox. Rupi rose, and everyone followed Sonjaa. Deka kept pace with them, leaving the rest of the group behind.

Rive, Kylac, and Friend followed from a distance. Kylac walked next to Friend. It felt so good to breathe another fox's scent, even one as unsettling as his.

"Are you better now?" Kylac said.

"No. But yes. The song is finally out of my head. I finished it. I brought Rel back—the equations are satisfied. My mind isn't constantly trying to work them out. I think... I think I'm in control again."

"Are you sure?"

Friend closed his eyes for a few steps. "No. For all I know, another equation is about to begin, and that will lead to others, and then more and more and more, all dependent on each other, all demanding to be solved at once, and I won't be able to stop." He faced the younger fox. "Kylac... I would have loved to work this out in a much better environment, but I couldn't. Deka will never forgive me. Will you?"

"Forgive yourself first," Kylac said. "I can imagine what you're going through, but it doesn't change the fifty-six worlds you destroyed."

Deka had run far ahead of them, trailing Sonjaa, Rupi, and the hatchlings, jumping around them, taking in their scents from all angles. She was walking everyone to the hub for this region. Deka ran after them, trying to tell her to wait.

"Would you like to know?" Friend asked.

Kylac met the ragged fox's eyes. "You mean—?"

"I figured out how to explain it. Maybe you'll be able to handle it better than Malim did."

Kylac's heart stopped and his mind looped. He forgot to breathe.

"I can help you," Friend said. "You wouldn't have to go through it alone. I wanted to take Malim with me, but he refused to leave Cham when he saw what he had done."

Kylac stared ahead, making a conscious effort to breathe. "How... How are we breathing?"

"Kylac..."

"We can't influence this world. Our feet don't touch the ground, our hands don't touch people, and we can't move objects. How can we affect the air? How can we see the light here? How can we smell anything? That's interacting with the world, too. How is this possible?"

"Do you want to know?"

"Yes," Kylac answered. "Yes, I want to know, but don't tell me. I don't want to end up like Malim."

"He couldn't handle it. I think you can."

"That's exactly what you told Malim, isn't it? Now he's dead. Why not tell Rive?"

The metallic raptor turned his muzzle and regarded Kylac. He could smell Kylac did not want to kill Friend, so he had not placed himself between them.

"I want somebody else to confirm what I've discovered," said Friend. "I don't think I can figure this out on my own. I need—"

Deka screeched from up ahead. They stopped and looked ahead. Sonjaa was gone, probably disappeared through a portal that wasn't there. The raptor began running back to meet them, killing claws up.

Kylac turned back to Friend. "What you need is to stop. This is time travel? I hope it was worth fifty-six planets."

He walked ahead and met Deka halfway. The raptor skidded to a halt, gripped Kylac by the shoulders and shook him around.

"I don't know where she went! I can't follow her! Kylac, help me think! Where is she?"

"I don't know where she is—I was with you, remember?"

"I know! I know..." Deka relaxed.

"I know where she is," Rive said. He and Friend had just now caught up to them. "That is, I know where she will end up."

3

Stephen and Norh were perched atop a mountain, looking down on the hub. It was below zero up here, but Stephen stood close to Norh, and his metabolism had sped up so high he didn't feel it. Below them lay the hub. Hundreds of people were in it, walking to and fro. Stephen recognized several species, even from this altitude.

"You still have to imagine this hub full of portals," Norh said, "but now you can see how the contacted universe used to be. Rel's hub is larger than most, but every planet had portals leading to other worlds. Every world had people coming and going. The people were free to go anywhere, anytime. If it rained, they simply took a sphere to a place where it was dry. If it was too hot, they took a way to somewhere cooler."

"Just like that?" Stephen said. "The entire universe just a brisk walk away?"

"Opens up many possibilities. On Hezgre, you said you didn't know where you would live if you had the choice. You have been isolated for so long you cannot even comprehend having the choice. Your mind needs its narrow perspective in order to make sense of reality. Having the possibilities open to you would overwhelm you."

"I wouldn't have to choose. I could live anywhere, any life I wanted, at any time. I could live on Kronia, then go to Hezgre when I want something different. Nobody's locked into a life on their own world. Nobody needs to be."

"If your kind had freedom, this would not be new to you."

"If Earth had a taste of this, I think it would be like when the Wall came down. Everyone broke out and never looked back."

"Your species wants to break out, but it has nowhere to go, and if it could leave, it lacks the maturity to survive anywhere else."

Stephen sat down, breathlessly watching the people come and go. "Just the idea... You can go anywhere you want. Live any way you want. You can just *go*."

"The things you can accomplish when you don't have to devote so much time to surviving. I realized this, and I tried to help many lone species realize it, too. One I remember very well lived on a dry planet. The carnivorous people scavenged the plants for water. Searching for succulent plants and hunting the animals that depended on those plants consumed their lives. They couldn't live anywhere else. Their bodies had adapted too well to the desert.

"I projected what kind of society they would form. Their competitive nature would drive the more aggressive ones to hoard the best plants, using thirst to keep the lesser ones down. Their society would develop into what you would call a monarchy. Perhaps they would devise a system of currency to ration the water. The ones with the most aggression would keep the more peaceful ones thirsty and needy to protect their own status, even when there was enough for everyone. I soared into the sky and observed how the waterways connected. There was a river far away from them, and I dug a new river into the desert for them."

"A whole river?"

"It would have been eight hundred miles. It took weeks to dig."

"Only weeks."

"They saw what I was doing, and at first they were puzzled and afraid, especially when I dug a lake. I connected the two halves, and the water flowed."

"What happened?"

"I had hoped that after spending so many generations searching for water, they would come to realize how much better their lives were when the water was plentiful and they did not have to scavenge for it anymore. Instead, the aggressive ones guarded the lake, and everyone else drank and hunted at their whim. If they didn't like someone, they let him die of thirst. Their minds were still too primitive to reason with. I tried to force them to cooperate, but it was no use. Years later I returned. Everyone was dead."

"The lesser ones fought back, didn't they?"

"It could have been a wonderful society."

Stephen watched the people on the hub, imagining the possibilities. "Why didn't you take them to another planet with another lone species? You could, couldn't you? Wouldn't that solve everything?"

"Each species is adapted to live on its own planet, for the food that grows there or the prey that lives there. Those people might have survived on another world, but their food would not have, and for a species so young, the shock of being taken somewhere they did not know how to survive would have killed them anyway. Others have tried it. Moving a species from its home world so late in its development is too much. By the time they are conscious enough to handle being moved, it is too late. We cannot change them even then."

Stephen sat in silence. He watched a few people disappear from the hub into imaginary portals, and then he laughed. "A new river. Just like that, you dug a new river. It would take a decade for us to do that, and we'd probably run out of money halfway through the project. But it still failed." Stephen traced another person's path along the hub. Then he turned up to Norh. "Does this happen to all Krone? Does everyone get the urge to go out and try to help a lone species?"

Norh's wings unfolded a little. "I am not unique among the Krone."

"So an entire population of dragons visiting uncontacted worlds, looking for people without a companion race, using their power to help them. An entire species realizing they can't change anything, so what's the point of having all that power?"

Norh was silent. Stephen continued.

"My wife would have said it's evidence of God. That the Krone were placed on that planet to help the Lost. But now the Lost are gone, so it's like the Krone have no purpose anymore."

Norh remained silent and still.

"Is this why the Krone hide in their caves for hundreds of years?" Stephen went on. "They realize that they may as well have died with their companion species because they really are nothing without them. Their power was only meant for the Lost. Is any of this right or am I just being insulting?"

Norh raised an arm and placed it on Stephen's back, wings unfolding halfway.

"You did not leave Earth simply because you wanted to know what else was out there. You begged the Relians to take you with them because you were hoping to learn something you could bring back that would change your world. If you found nothing, you had planned to run away and hide at the first opportunity."

Stephen felt like his mind was an open book, and everyone could read it but himself.

"Yeah. Stupid. I see that now. I like to think anybody would have done the same thing. Every human being secretly hopes to change the world. Then we grow up, and we find we're totally powerless. As much as I've seen, as much as I've been through in just a few short months... There's nothing I can just tell anyone. Nothing. Nobody will believe me, and nobody will understand."

Norh wrapped his fingers around Stephen, holding him as close as possible without crushing him. "Where would you have hidden?"

"Eiae. That was the first place I felt I was actually starting to get used to. I was hoping to get lost there and nobody would find me. Or maybe I'd go back to Uiv. I'd force myself to get to know those people, and this time I wouldn't be scared."

"You remind me so much of myself when I was young. You can't save your planet, Stephen, but you can save yourself."

Stephen smiled, reached across his chest and held one of Norh's claws. Norh squeezed him, just a little, unfolding his wings the rest of the way.

"Deka and Kylac have caught up to Rive and Friend," Norh said. "They chose to spare their lives, so there is no reason to stay with them. You are ready to come with me now. We will part ways with Deka and Kylac when we meet them next. I have already been working on a way to Pryip. I will let you experience as much as you can. Eventually the portals will reopen, and you can live any life you choose. With enough time, I am sure everything will make sense."

Stephen laughed. "Maybe it will."

Norh gazed out over the hub. Stephen watched, too, the Krone's hand still clutching him.

4

It had been so long since Kylac and Deka had been to the hunting grounds of Rel. An entire continent reserved for prey to roam and multiply freely. Rel had hundreds of portals open across it for carnivores of all planets to come and hunt as they pleased.

None of the animals were running now. The herds lay still, huddled into themselves, shivering even as carnivores from all over the contacted universe stalked them. The animals were placid, as if they had come to accept their fate and resolved to meet it with dignity. Many carnivores acted bewildered, though some took advantage of the easy kills.

Relian canines generally ate of the kills their raptors made. It was rare for a fox to hunt, so the canine out in the field stood out. Friend's past self was on all fours, charging a pack of four-legged bull-like creatures. They weren't fleeing for their lives, so Friend landed on one and easily clamped his jaws around its neck.

The Archeons from the future watched past-Friend take down the animal from a distance. They dared not get too close to any packs of animals, or to anyone hunting. Past-Friend stood triumphant on top of the body, howling in victory. His scent projected itself far from the kill, and he smelled as though he were wallowing in the old ways. Rive's past self approached Friend, which brought the fox's scent back down, and they shared the kill.

One by one the Archeons sat and waited. Deka remained standing, scenting the air constantly.

The tailless Friend had sat down beside Kylac, and Kylac looked at Friend from time to time, pondering the tremendous thing he had offered. Kylac felt privileged Friend was asking him and not Rive, and he wanted to believe he was capable of controlling it—but every time he thought that, he thought of Malim's body torn apart on the edge of his world. Kylac thought of Deka. If something were to happen to him, what would that do to Deka? Kylac turned to his raptor, who stood just ahead, looking around, eager to run.

Kylac was sure Deka was right; this should not be explored, not if it meant entire planets had to be destroyed, and the only reward was a glimpse into a past that could not be changed. Kylac yearned for the fantasy worlds of the movies Stephen owned. Humans were so desperate to have influence over their world, and the society in which they felt trapped, that they had created the idea of changing what had already happened. The reality of the situation was that everything had already happened, and there was no way even to interact with it.

Walking around the planet, unable to affect anything, had been eerie. They passed hundreds of people unseen and unheard, unable to touch or influence anyone. It was also dangerous because they could be trampled by something as small as a child, so they constantly had to dodge people. Resting anywhere was dangerous. This spot seemed to be free of all activity though, and they were glad for the chance to sit.

Deka suddenly perked up. Sonjaa and Rupi were walking through the hunting grounds together, headed straight for the Rive and Friend of the past. Kylac stood, and so did the metal Rive and the tailless Friend.

"I asked her to meet me," present-Friend said.

Deka whipped his head around and glared at him. "Why?"

"I'd been trying to talk to both of you about this, remember? You would not listen to me. I was out of Archeons to talk to. Sonjaa always listened. I thought she would want to hear this."

"Hear what?"

"Hear the idea I had. She wasn't an Archeon, but—"

"There are a thousand Archeons in the contacted universe!" Deka said, baring his teeth. "Why not them? Why did you want to talk to my mate?"

"Because she listened to me," Friend said. "Everyone else was closed to the idea."

Deka turned back to the scene playing out before them. Sonjaa and Rupi had just caught up to the past Rive and Friend. Other raptors and a few offworld predators gathered around the fox's kill, and even a few other foxes.

Deka ran to the kill as well. Kylac rose to all fours and chased his raptor. The metal Rive looked at present-Friend. He wasn't moving, so Rive didn't move either.

Kylac caught up to Deka a few paces from the kill—close enough to hear everyone, but not so close they would risk being caught in a crowd they couldn't escape.

Kylac observed his surroundings. This felt familiar.

Deka met his eyes. He agreed.

The raptors were feeding their foxes. Sonjaa was pulling strips of muscle off the large animal and feeding them to Rupi. Her fox was pregnant, so Sonjaa was practically stuffing her with food. Deka smiled with his hands.

Past-Friend had obviously eaten his fill, and now he approached her. The fur around his muzzle was still covered in blood, and his chest and arms were also splattered with it.

"Sonjaa?"

"Friend, if this is about—"

"It is, but—"

"Deka told me all about it. I don't know why you keep going back to it. It's a waste of your time."

"The idea is sound," past-Friend said. "Archeons only ever make ways across one dimension, but what if there are other directions to make a way?"

"Friend, find someone else to bounce ideas off of. You've been talking about it for years. I'm tired of it."

"I had a new idea," he said. "It's about motion."

"I know, you already told me. The universe is in motion, everything is moving, stop the motion and you stop time."

"Time leaves a trail."

Sonjaa was silent. For the first time she seemed to be listening.

"Take your finger, drag it across the surface of a lake. It leaves a trail in the water. If you could stop time at one instant, you would see an impression of your finger in any location. The universe is doing the same thing as it moves through... wherever it is. This means it is possible to create a way back through the motion of the universe. If someone can figure out where the universe is located, and where it's moving, it's possible to create a portal back to where it's been. Maybe even anticipate where it's going and make a way into the future."

"But the water always comes back and fills the gap."

"On the scale of the whole universe, it can take millions of years for that to happen."

"There is no past to go back to, and you can't go into the future. It hasn't happened yet."

"What if the finger has already moved across the whole lake? What if everything already is, and we're just moving on the path the universe already carved through the water?"

Sonjaa licked blood from her lips. "It's an idea. So if everything has already happened, why would the universe need to move?"

Deka was snarling at Friend's past self. "Sonjaa, get out! Get away from him! Find a portal offworld now!"

Past-Friend and Sonjaa were walking away from the kill. Deka kept pace with his mate, screaming in her ear. He then began clawing and scratching at past-Friend. His claws never touched the fox. A few dozen paces away from the kill, Deka finally tired and settled into walking just behind them, beside Kylac.

Rive and the tailless Friend joined them from behind. The metal theropod walked with Deka, his artificial skin and limbs flexing and whining. They had no joints or any obvious flexing points; the metal seemed to distort somehow. The Friend of the present walked next to Kylac.

"It doesn't make sense," Sonjaa said. "Is there a future or isn't there? Is time a perception, or is it a physical place?"

"Time is the result of the universe moving from where it was to where it is," past-Friend said. "It does leave impressions behind. Traveling to the future is not possible in the same way as the past because the future has not happened yet, but it is possible to create a way into a time and place where the universe will be."

"An impression," said Kylac, turning to face the other Friend. "You already knew."

"I knew it as a concept," said the tailless fox. "I had no idea it would actually happen."

"So where is the universe?" asked Sonjaa.

"That is the big mystery," said past-Friend. "As an Archeon, I make ways into this part of the universe. Only around the finger, the present, but what if I could reach backwards? What if there is no time? Only a different place?"

"It would only add another variable," said Sonjaa. "Identifying where on the trail you want to go, and then deciding on a location in it."

"That could be it," said the past-Friend.

Deka was still snarling. "Sonjaa, run! Don't listen to him!"

Kylac glanced at the tailless fox beside him. "How soon?"

"Just a few breaths."

"Why are you here?"

"To remember how it began. I actually can't recall this moment."

Kylac turned to Sonjaa and past-Friend again. They were still walking away from the kill across the open field. Friend stumbled, and Sonjaa caught him just before he fell.

"What's wrong?" she said.

"I just thought of something! Yes! It's— It's—"

The group of time travelers halted a few paces from the scene and watched. The tailless Friend finished the sentence his past self could not complete.

"It's just as when I was an apprentice. After meditating for so long, finally you become aware of the universe, and your mind opens its first way."

Past-Friend was holding his head, panting and whining. Sonjaa held him, forcing him to meet her eyes.

"What's wrong? Friend, are you all right?"

"It makes sense..." he said. "It makes sense!"

Deka backed away. So did Kylac and the other two.

Friend screamed. Sonjaa let him drop to the ground. The ground shook. A few paces away, just above the ground, a tiny antisphere hovered. The ground beneath it was puckering, like water about to boil. Sonjaa stood up.

"Sonjaa, run!" Deka screamed.

"That's it!" said past-Friend. "The past! I don't know where, but I think I'm in touch with it! Sonjaa!" Friend rolled over and stood on one knee, gazing at it. "It leads outside the universe! That's a way through time!"

People were running from all over the hunting grounds to see the new sphere. It did not resemble any sphere they had ever seen before. They pushed and shoved and bumped Deka, Kylac, the metal Rive, and the tailless Friend around as they gathered. The visitors from the future couldn't fight the people or move them or touch them in any way, and quickly they became trapped within the inadequate space between dozens of raptors and foxes and offworlders.

The antisphere was surprisingly stable and calm. Past-Friend stood up, approached it, never taking his eyes off it.

"No..." he said. "No, something is missing. Something is wrong. Where is this?"

"What do you mean?" Sonjaa said.

"This way... It doesn't lead anywhere. So where does it go? Where am I? Where is this?"

"Friend—" Sonjaa walked to him, grabbed his muzzle and held it to hers. "Tell me what that thing is!"

"I don't— I don't know! I thought I saw down the trail in the lake! It's supposed to be! It's what I was thinking, but I don't know where it goes! This doesn't— This doesn't— This is—" Friend screamed.

The antisphere began expanding. The ground shook harder, the grass underneath it ripped up and disappeared into it.

Sonjaa held past-Friend harder. "Stop! Friend, stop it!"

"This is— This is— This is— Where am I?! Where does it go?!"

The antisphere intersected the ground. The ground shook as soon as it touched. It seemed to hurt Friend as well, for he screamed louder.

"You can stop this!" Sonjaa shouted over the fox's voice, shaking him around. "Back off! Close the portal!"

"I can't! I don't know how I started it!"

The antisphere expanded. Everything shook. Parts of the planet that weren't near the antisphere pulled up and fell into it.

Deka and Kylac struggled to free themselves, but none of the people were moving. They stared ahead, gazing into pure nothing. Rive was trying to free himself as well. Only the tailless, sleep-deprived Friend remained calm, squished between two raptors and five foxes.

Sonjaa still held past-Friend, shaking him, yelling at him to close it and pull his mind out of wherever it was, but the fox was barely coherent, and his words were lost in the planet quaking under their feet.

Past-Friend suddenly came back to himself, tore from Sonjaa's grip, and stumbled away.

Deka snarled, still trying to climb over the statues of the people. "Run!"

"I can't stop it!" screamed the Friend of the past. "I don't know where I am! I don't know where it goes! I didn't tell it to spin like this! I don't understand!"

"Stop thinking about it!" Sonjaa shouted.

The antisphere had grown five times larger, towering over everyone, grinding up the ground. It was less than a pace from her. Sonjaa turned her head and faced it, and then she lifted off her feet and slipped inside.

At that moment, the observers ceased being statues. Suddenly Kylac and Deka were actually touching them. The metal Rive and present-Friend felt the same effect.

Sonjaa twirled around inside the antisphere. The past-Friend took off running on all fours back to the kill, where the non-metal Rive was staring. Sonjaa held her arms and tail out and managed to stabilize herself.

She saw the congregation from within the growing antisphere. Her eyes locked with Deka's.

"Deka, help me!" she shouted, her voice quieter than it should have been.

The raptor began knocking people down, but even when they fell, they made a pile in front of Deka, and he still could not move. Sonjaa began spinning around again, moving closer and closer to the center of the antisphere.

"Deka, what's happening?!"

She stopped, looking over the rest of the spectators. The metal Rive had managed to push down enough people to escape. The tailless Friend stood still, watching. Sonjaa began to spread across the sphere.

"Friend! Rive! Help me, please! Someone! Help!"

Sonjaa dissipated across the surface of the antisphere, and it became dark again. The sphere continued expanding. Columns of soil, mantle, and core rose out of the ground, taking columns of tissue and bone with them.

Deka crouched and leaped over the crowd. Kylac climbed over the hypnotized raptors and foxes and touched down outside them, and he ran with Deka.

Rive had made a path through the people easily with his metal arm and was scenting for Friend. The tailless fox wiggled between a pair of raptor legs just a moment later, and they ran side by side.

Their portal to the other side of the planet was in sight. Deka and Kylac lined themselves up and ran straight through. They emerged where Sonjaa had been sleeping. It was much calmer here, though the rumbling through the ground reminded them it was not over.

Kylac ended the portal leading to the hunting grounds, and half a breath later Rive and Friend poured out of their sphere, and it closed behind them. Deka snarled at Friend, but Kylac grabbed him, and as a group they ran for the nearby portal that led to the hub.

The hub was in a much greater panic, as the anti-sphere loomed over the horizon, and the ground shook harder than ever. The four of them dashed through the hub, shoving people out of the way, climbing over fallen people.

Rive tried to pull a few people to the portal with him, but they were staring at the antisphere as it tore up the horizon, taking larger and larger columns of ground from beneath everyone's feet. It lifted entire people and tore them apart in midair.

The metal raptor gave up and left the people where they stood. He ran through the way to Vico. Deka, Kylac, and Friend ran through it as well, emerging in the thin atmosphere and weak gravity of Rel's smallest moon.

The view was as cinematic as one of Stephen's movies. The planet had another planet growing inside of it, but this world pulled pieces of Rel apart as it swallowed what it didn't destroy. The oceans disappeared into it. The mantle and core spilled out and entered a brief orbit around the planet before being pulled into the antisphere as well.

The antisphere was now twice the size of Rel, and finally the entire planet was gone. The antisphere lingered for a few moments, and then it dissipated from the inside out.

Deka glared at Friend. He raised his killing claws, spread his hands, and charged, making long strides in the weak gravity. This time Rive did not leap to intercept, but seemed lost in thought. Friend stood still. Deka opened his mouth.

A large portal opened between Friend and Deka. Deka gasped, pulled to the side and swerved just enough to avoid it. He skidded across the ground, skipping and bouncing as though he were full of helium. Finally he settled on his side.

A hundred portals opened up, all leading offworld, each one uniform in size and resting less than a claw's reach from one another. The portals surrounded Deka, Rive, and Kylac, closing them in, and Friend stood in the middle, the only one with room to move.

Deka screamed as he got up. "Friend!"

"I solved the equation," said the tailless fox. "I can think again. Finally, I can think! After you spend a year trying to calculate a way through time, figuring out traditional ways becomes so much simpler. I calculated all these offworld portals just now."

Kylac jumped carefully, taking a closer look at Friend. He still stood in place, calm, collected, at ease for the first time. Deka jumped, but not too high. Rive stood, gazing up at the sky where Rel used to be.

"I don't need days to figure out a way offworld. I have seen the universe as a whole! I saw where everything is!"

Deka wanted to leap over the spheres and attack, but he did not trust he could land without cutting himself between two or three spheres. Instead he stood in his prison and snarled.

"This needs to be understood!" Friend shouted. "The entire universe is the same world to me! I know my mistake! It's not about time travel! My original idea—the one that started it all was pondering where the universe is! Where it's going! Why it's going anywhere! I need to understand what the Lake is made of!"

Kylac's ears folded back. He whimpered for the loss of Rel, and he thought of Stephen and Norh, wondering where they were and if they made it back.

Suddenly every portal between him and Friend winked away, opening a clean path to the tailless fox. Friend charged Kylac. The fox stood up and had just enough time to open his mouth when Friend slammed into him and threw Kylac into the sphere behind him.

The spheres closed, leaving just Rive's spinning portals giving Vico its temporary atmosphere. The two raptors stood alone on a desolate moon.

Rive ran toward Deka.

"Stay away from me!" Deka snapped.

"Deka, there's something you should know."

"What? What should I know? Why did you stop me? Why didn't you have the guts to kill Friend after the first planet? Why did you let this happen?"

Rive huddled into himself, avoiding eye contact. "Because he's my fox."

Deka reached out and slashed him across the face, but his claws only sang as they swiped off the metal surface. Deka turned away, stumbling in the poor gravity.

"All you had to do was let me kill him! This would be over, and this never would have happened if you'd had the fucking courage to end it!"

"I can't kill my fox."

"But you wanted someone to! Now Friend has my fox. And Sonjaa... What happened to her?"

"Deka, we found survivors."

Deka turned, his breath caught in his throat. The grey and tan raptor of flesh and metal was standing up straight now, facing him.

Deka took a breath. "Relians?"

"One hundred and five of them. I've been moving them to one world. That's why Friend was so determined to restore Rel. He wanted to give it back to them. I wanted him to be right. That's why I didn't kill him."

"Where are they?"

"I'll make the way. I've been working on it since we set foot on Rel. It won't take long to adjust to Vico."

Deka's killing claws lowered as he regarded the metal raptor. He sat down and waited. Rive sat three paces away.

Rive turned to the ground and began shrieking softly, like a hatchling. Deka couldn't tell if Rive was crying for his fox, for the planets they lost, or for himself.

Deka did not speak. He had run out of words.

About the Author

James L. Steele has had the idea for the Archeon series in his head since the mid-1990s.

He has been published in various anthologies and magazines, including: *Solarcide, Allasso, Different Worlds, Different Skins: V.2, Tall Tales with Short Cocks V.2, Bourbon Penn, Gods with Fur, Claw the Way to Victory*, and *Fictionvale*.

His sci-fi novel *Huvek* is published through Argyll Productions.

He lives in Ohio, where he pursues his hobby of becoming a wine connoisseur while having between three and nine existential crises per day.

Website: JamesLSteele.com

Blog: DaydreamingInText.blogspot.com

Twitter: @JLSteeleAuthor